REVIEW OF THE RESEARCH PROGRAM OF THE PARTNERSHIP FOR A NEW GENERATION OF VEHICLES

SIXTH REPORT

Standing Committee to Review the Research Program of the
Partnership for a New Generation of Vehicles

Board on Energy and Environmental Systems
Commission on Engineering and Technical Systems
Transportation Research Board

NATIONAL ACADEMY PRESS
Washington, D.C.

National Academy Press • 2101 Constitution Avenue, N.W. • Washington, D.C. 20418

NOTICE: The project that is the subject of this report was approved by the Governing Board of the National Research Council, whose members are drawn from the councils of the National Academy of Sciences, the National Academy of Engineering, and the Institute of Medicine. The members of the committee responsible for the report were chosen for their special competences and with regard for appropriate balance.

This report has been reviewed by a group other than the authors according to procedures approved by a Report Review Committee consisting of members of the National Academy of Sciences, the National Academy of Engineering, and the Institute of Medicine.

This report and the study on which it is based were supported by Contract No. DTNH22-94-G-07414. Any opinions, findings, conclusions, or recommendations expressed in this publication are those of the author(s) and do not necessarily reflect the view of the organizations or agencies that provided support for the project.

Library of Congress Card Number: 00-104666
International Standard Book Number: 0-309-07094-5

Available in limited supply from:
Board on Energy and Environmental Systems
National Research Council
2101 Constitution Avenue, N.W.
HA-270
Washington, DC 20418
202-334-3344

Additional copies are available for sale from:
National Academy Press
2101 Constitution Avenue, N.W.
Box 285
Washington, DC 20055
800-624-6242 or 202-334-3313 (in the Washington metropolitan area)
http://www.nap.edu

THE NATIONAL ACADEMIES

National Academy of Sciences
National Academy of Engineering
Institute of Medicine
National Research Council

The **National Academy of Sciences** is a private, nonprofit, self-perpetuating society of distinguished scholars engaged in scientific and engineering research, dedicated to the furtherance of science and technology and to their use for the general welfare. Upon the authority of the charter granted to it by the Congress in 1863, the Academy has a mandate that requires it to advise the federal government on scientific and technical matters. Dr. Bruce M. Alberts is president of the National Academy of Sciences.

The **National Academy of Engineering** was established in 1964, under the charter of the National Academy of Sciences, as a parallel organization of outstanding engineers. It is autonomous in its administration and in the selection of its members, sharing with the National Academy of Sciences the responsibility for advising the federal government. The National Academy of Engineering also sponsors engineering programs aimed at meeting national needs, encourages education and research, and recognizes the superior achievements of engineers. Dr. William A. Wulf is president of the National Academy of Engineering.

The **Institute of Medicine** was established in 1970 by the National Academy of Sciences to secure the services of eminent members of appropriate professions in the examination of policy matters pertaining to the health of the public. The Institute acts under the responsibility given to the National Academy of Sciences by its congressional charter to be an adviser to the federal government and, upon its own initiative, to identify issues of medical care, research, and education. Dr. Kenneth I. Shine is president of the Institute of Medicine.

The **National Research Council** was organized by the National Academy of Sciences in 1916 to associate the broad community of science and technology with the Academy's purposes of furthering knowledge and advising the federal government. Functioning in accordance with general policies determined by the Academy, the Council has become the principal operating agency of both the National Academy of Sciences and the National Academy of Engineering in providing services to the government, the public, and the scientific and engineering communities. The Council is administered jointly by both Academies and the Institute of Medicine. Dr. Bruce M. Alberts and Dr. William A. Wulf are chairman and vice chairman, respectively, of the National Research Council.

Project Staff

JAMES ZUCCHETTO, director, Board on Energy and Environmental Systems
(BEES)
NAN HUMPHREY, senior program officer, Transportation Research Board
SUSANNA E. CLARENDON, senior project assistant and financial associate,
(BEES)
CAROL R. ARENBERG, editor, Commission on Engineering and Technical
Systems

Acknowledgments

The committee wishes to thank all of the members of the Partnership for a New Generation of Vehicles who contributed significantly of their time and effort to this National Research Council (NRC) study, either by giving presentations at meetings, responding to requests for information, or hosting site visits. The committee also acknowledges the valuable contributions of other organizations outside that provided information on advanced vehicle technologies and development initiatives. Finally, the chairman wishes to recognize the committee members and the staff of the NRC Board on Energy and Environmental Systems for their hard work organizing and planning committee meetings and their individual efforts in gathering information and writing sections of the report.

This report has been reviewed by individuals chosen for their diverse perspectives and technical expertise, in accordance with procedures approved by the NRC's Report Review Committee. The purpose of this independent review is to provide candid and critical comments that will assist the authors and the NRC in making the published report as sound as possible and to ensure that the report meets institutional standards for objectivity, evidence, and responsiveness to the study charge. The content of the review comments and draft manuscript remain confidential to protect the integrity of the deliberative process. We wish to thank the follow individuals for their participation in the review of this report: Gary Byrd, consulting engineer; Tom Cackett, California Air Resources Board; David Holloway, University of Maryland; Patrick Flynn, Cummins Engine Company, Inc.; John P. McTague, Ford Motor Company (retired); Phillip S. Myers, University of Wisconsin; Jerome G. Rivard, Global Technology and Business Development; Robert J. Schultz, General Motors Corporation (retired); F. Stan Settles,

University of Southern California; Dale F. Stein, Michigan Technological University (retired).

While the individuals listed above have provided constructive comments and suggestions, responsibility for the final content of this report rests solely with the authoring committee and the NRC.

Contents

EXECUTIVE SUMMARY   1

1   INTRODUCTION   13
    Program Milestones, 15
    Scope of Review, 16

2   DEVELOPMENT OF VEHICLE SUBSYSTEMS   18
    Candidate Systems, 18
    Internal Combustion Reciprocating Engines, 19
    Fuel Cells, 31
    Electrochemical Energy Storage, 39
    Power Electronics and Electrical Systems, 43
    Structural Materials and Safety, 46

3   SYSTEMS ANALYSIS   56
    Program Status, 56
    Assessment of the Program, 57

4   CONCEPT VEHICLES   60
    General Motors, 61
    Ford, 65
    DaimlerChrysler, 66
    Summary, 68

5   MAJOR CROSSCUTTING ISSUES   69
     Background, 69
     Major Achievements and Technical Barriers, 71
     Adequacy and Balance of the PNGV Program, 75
     Fuel Economy and Emission Trade-offs, 81
     Fuel Issues, 83

6   PNGV'S RESPONSE TO THE FIFTH REPORT   88

REFERENCES   91

APPENDICES
     A   Biographical Sketches of Committee Members, 97
     B   Letter from PNGV, 104
     C   Presentations and Committee Activities, 108
     D   United States Council for Automotive Research Consortia, 111

ACRONYMS AND ABBREVIATIONS   113

Tables and Figures

TABLES

2-1   Design Targets and Current Performance for Short-Term Energy Storage, 40
2-2   Current Specifications and Target Specifications for Power Electronics, 44
2-3   Weight-Reduction Targets for the Goal 3 Vehicle, 47
2-4   Material Cost Targets, 52

4-1   Comparative Attributes of PNGV Concept Vehicles, 62
4-2   Comparison of the Toyota and Honda Hybrid Vehicles, 64

5-1   Infrastructure Investment for the Production and Distribution of Hydro-
       gen and Methanol, 86

FIGURES

2-1   Alternatives for energy conversion, 21

4-1   The General Motors Precept concept vehicle, 65
4-2   The Ford Prodigy concept vehicle, 66
4-3   The DaimlerChrysler ESX3 concept vehicle, 67

5-1   Distribution of DOE's Office of Advanced Automotive Technologies
       budget for PNGV (by technology), 76

REVIEW OF THE RESEARCH PROGRAM
OF THE
PARTNERSHIP FOR A
NEW GENERATION OF VEHICLES

SIXTH REPORT

Executive Summary

This is the sixth report by the National Research Council Standing Committee to Review the Research Program of the Partnership for a New Generation of Vehicles (PNGV). The PNGV program is a cooperative research and development (R&D) program between the federal government and the United States Council for Automotive Research (USCAR), whose members are DaimlerChrysler Corporation, Ford Motor Company, and General Motors Corporation (GM). A major objective of the PNGV program, referred to as the Goal 3 objective,[1] is to develop technologies for a new generation of vehicles with fuel economies up to three times (80 miles per gallon [mpg]) those of comparable 1994 family sedans. At the same time, these vehicles must be comparable in terms of performance, size, utility, and cost of ownership and operation and must meet or exceed federal safety and emissions requirements. The intent of the PNGV program is to develop concept vehicles by 2000 and production prototype vehicles by 2004.

In this report, the committee continues to examine the overall adequacy and balance of the PNGV research program to meet the program goals and requirements (i.e., technical objectives, schedules, and rates of progress). The committee also discusses ongoing research on fuels, propulsion engines, and emission controls to meet emission requirements and reviews the USCAR partners' progress on PNGV concept vehicles for 2000.

---

[1] Goal 1 is to improve national competitiveness in manufacturing significantly. Goal 2 is to implement commercially viable innovations from ongoing research on conventional vehicles. Because the Goal 3 concept vehicle demonstrations are focused on relatively near-term technologies, the distinctions between goals 1, 2, and 3 are becoming blurred. Nevertheless, the committee tried to focus on PNGV's efforts to meet Goal 3.

PROGRESS AND MAJOR ACHIEVEMENTS

Considering the magnitude of the challenges facing the program, PNGV is making good progress. As the program has evolved, the PNGV technical teams have become more effective and have developed good working relationships. In addition, the USCAR partners have created substantial vehicle engineering teams devoted to the development of the concept vehicles, which were unveiled in January and February 2000. PNGV has also responded positively to most of the committee's recommendations in the fifth report. In general, the committee congratulates the PNGV partners on their progress in the past year.

Concept Vehicle Milestone

The PNGV has met its 2000 concept vehicle schedule and milestone with the unveiling, in January and February 2000, of the USCAR partners' concept vehicles (DaimlerChrysler's ESX3, Ford's Prodigy, and GM's Precept). Meeting this milestone represents an outstanding industry effort. The concept vehicles are intended to establish the functional benefits of their designs but may include components for which validated manufacturing processes and affordable costs have not yet been demonstrated. As expected, each manufacturer has taken a somewhat different approach, but the concept cars all share technology and know-how developed in PNGV, some of which is finding its way into current production vehicles (as called for in Goal 2).

All of the concept vehicles incorporate hybrid-electric drive trains designed around small, turbocharged, compression-ignition, direct-injection (CIDI or diesel) engines that shut down when the vehicles come to rest. These hybrid electric vehicles (HEVs) have been constructed according to sophisticated structural optimization techniques with high strength-to-weight materials, such as aluminum and composites, in both bodies and interiors. Every aspect of these cars, including wheels, tires, interior components, front, back, and side windows, rear vision devices, and aerodynamic drag, has been designed to reduce weight and increase efficiency. Friction has been reduced in almost every rotating component. All three cars are expected to achieve 70–80 mpg (gasoline equivalent), the ESX3 at 72 mpg, the Prodigy at 70 mpg, and the Precept at 80 mpg, although tests have not been run to confirm these figures. Emissions are targeted at the Environmental Protection Agency's (EPA's) Tier 2 standards, but the after-treatment systems required to achieve these standards have not been defined.

In summary, the concept cars represent a major milestone toward meeting the PNGV Goal 3, and each contributes significantly to our understanding of the challenges this goal represents. In addition, GM has built a fuel-cell version of the Precept, which is packaged in the same basic chassis and body and includes a hydride hydrogen storage system. The fuel-cell Precept is expected to achieve a gasoline-equivalent fuel economy of more than 100 mpg when supplied with hydrogen. A fully functional version of this car is expected by the end of 2000.

The automotive companies have taken different approaches to meeting Goal 3. For example, the Ford Prodigy power train system is considerably simpler than the power train in the GM Precept. Thus, the Ford system is closer to meeting the affordability target of Goal 3 but sacrifices some fuel economy by limiting the amount of potential regenerative braking. The DaimlerChrysler ESX3 has large injection-molded plastic body sections, which have the potential for building a body structure that both weighs less (the ESX3 curb weight is lower than that of the Precept or the Prodigy) and costs less than conventional steel bodies and is completely recyclable. The aluminum body construction used for the Ford Prodigy and the GM Precept currently costs significantly more than a comparable steel body, but as fabrication of aluminum bodies improves, they may become competitive with steel.

Goals 1 and 2

Although most of the discussion in this report about achievements and barriers is focused on Goal 3, continuing and significant progress has also been made toward achieving goals 1 and 2. For example, a project has been successfully completed demonstrating continuous cast sheets of Series 5000 aluminum for body structures, and a follow-up project to develop similar processes for exterior body parts is under way. Several smaller efforts to expand aluminum manufacturing and assembly capabilities are in progress, and an alliance between the automotive and aluminum industries has been formed to address standardization, scrap recovery, and other issues. Cost reductions, improved properties, and new manufacturing techniques for carbon-fiber composites, as well as the recycling and design of hybrid material bodies, have also been achieved. Aluminum springback predictability techniques have also been developed.

Goal 3 Achievements

Substantial technical progress has been made in reducing the energy requirements for propelling the vehicle (e.g., reduced mass, drag, etc.) and for supplying auxiliary loads (e.g., heating, air conditioning, etc.). Continual improvements have been made in the efficiency and performance of power plants (four-stroke direct-injection engines, fuel cells [as a long-range technology]), energy storage (batteries) for HEVs, and the development of modeling and simulation techniques. The three concept vehicles represent the results to date of these substantial efforts by the USCAR partners.

Vehicle Engineering and Structural Materials

A number of accomplishments have been made in vehicle engineering, including the fabrication and testing of a lightweight, hybrid-material body to validate a weight reduction of more than 40 percent; the completion of an energy-efficient, occupant comfort project with a 75-percent reduction in required energy; and the completion of a lightweight interior project demonstrating a 157-pound reduction in interior weight. In addition, projects have been initiated or continued on a high (42 percent) payload/curb-weight ratio, low rolling resistance (run-flat) tire, underbody airflow management, and energy-efficient side windows.

The committee believes that, as the PNGV program moves toward the development of 2004 production-prototype vehicles, affordability will be a key requirement. PNGV should closely monitor the development of an efficiently designed and fabricated steel-intensive vehicle being developed by the Ultralight Steel Auto Body Consortium (which is not part of PNGV), including the possibility of a hybrid steel-aluminum vehicle. In addition to continuous casting of aluminum sheet, PNGV has also made progress in developing lower cost aluminum and magnesium casting processes, lower cost powder-metal processes for aluminum-metal matrix composites, and a microwave process for lower cost carbon fiber.

Four-Stroke Direct-Injection Engines and Fuels

A number of new collaborative projects in advanced combustion and emission controls have been initiated; other projects are continuing to advance the understanding of catalysis, as well as to define fuels issues. New catalysts have been developed: a catalyst with lower "light-off" temperature and better nitrogen oxides (NO_x) reduction; a zeolite-supported catalyst to improve NO_x reduction; and microporous catalysts that show promising results.

Other projects are exploring novel means of reducing emissions of particulate matter (PM), improving the effectiveness of exhaust gas recirculation, and gaining a better understanding of combustion processes. A PM filter has been developed that has the potential to remove up to 90 percent of diesel particulates and can be regenerated at idle using microwave heating techniques. In addition, a plasma-assisted catalyst has been developed that shows high NO_x conversion rates even in the presence of sulfur.

It has also been demonstrated that changes in fuel formulation could reduce diesel PM emissions by 50 percent and NO_x emissions by 10 percent. Refinery models are being used to evaluate the effects of various formulations on diesel fuel costs and to define other issues related to fuels infrastructure.

Fuel Cells

Both national laboratories and industry contractors have made notable accomplishments on fuel cells. A gasoline (fuel-flexible) fuel processor has been operated in conjunction with a 10-kW proton exchange membrane (PEM) stack, but not as an integrated system. Microchannel fuel processing of iso-octane has been demonstrated and a new higher temperature nonair-sensitive fuel processing catalyst has been developed. A fuel anode that is much more tolerant to carbon monoxide has been demonstrated and a high power density fuel cell stack, low-cost composite bipolar plates, and low-cost membrane electrode assemblies have all been demonstrated by industry. Improvements in modeling and simulation are continuing, and production techniques for low-cost molded bipolar plates have been demonstrated.

Batteries

Research and development on batteries have been focused primarily on nickel metal hydride, lithium-ion, and lithium-polymer batteries because of their potential for high power-to-weight ratios for HEV applications, and (at least in some cases) low cost. During the past year, PNGV has received and evaluated a 50-V nickel metal hydride battery module and has received four lithium-ion battery modules for testing. The development of a 300-V battery system has been initiated. Lithium-ion electrochemistries projected to increase calendar life from two years to three to five years have been identified. Failure mechanisms and abuse tolerance issues for lithium-ion systems are better understood. However, projected costs are three times higher than the PNGV goals.

Nickel metal hydride batteries were used in Ford's Prodigy concept vehicle; lithium-ion batteries were used in DaimlerChrysler's ESX3 concept vehicle; and nickel metal hydride (and later lithium polymer) batteries were used in GM's Precept concept vehicle.

Power Electronics and Electrical Systems

The committee compliments PNGV's comprehensive power electronics and electrical systems program for its excellent organization and management. The program has accomplished a good deal in the past year. For example, improved direct-current power bus capacitors suggest that improved performance and reduced cost are feasible. An ongoing project is helping to assess the mechanical reliability of electronic ceramic devices and less expensive alternatives through mechanical characterization. Processes to fabricate neodymium-iron permanent magnets with up to 25 percent higher strength than current magnets are being developed, and a facility has been completed and work initiated on characterization of the magnets. In addition, a 100-kW inverter with a power density of 11 kW/kg has been developed, and collaborative efforts are under way on a lower cost 100-kW motor controller.

MAJOR BARRIERS

In spite of substantial accomplishments in virtually every technical area of the PNGV program, formidable barriers remain to be overcome. The realization of an advanced high fuel economy vehicle that meets Goal 3 requirements and is acceptable in the marketplace still faces significant barriers of cost, emissions, and fuel infrastructure. New business arrangements, such as the Daimler-Chrysler merger and the Delphi spinoff from GM, as well as the fact that the program has moved into the production prototype development stage, have complicated reaching a consensus on precompetitive projects.

Cost Challenge

As the committee has noted in previous reviews, vehicles incorporating both near-term and long-term technologies face critical cost barriers. In the committee's opinion, meeting the three most critical requirements (emissions, fuel economy, and cost) of Goal 3 by 2004 is very unlikely. Although 80-mpg fuel economy appears to be technically feasible, the cost requirement is clearly unattainable. The Tier 2 emissions standard appears to be attainable only with a fuel economy well below 80 mpg, and, even then, will be difficult to achieve in production vehicles with adequate probability for meeting the certification period of 100,000 miles. The development of vehicles of radical design (e.g., a fuel-cell vehicle) for mass production that meets all of the Goal 3 objectives by 2004 is also highly optimistic.

High cost is a serious problem in almost every area of the PNGV program, and the costs of most components are higher than their target values. For example, neither aluminum nor composite materials are yet projected to reach costs competitive with steel for most major vehicle components; the CIDI engine will require low-cost after-treatment and cost reduction for common-rail fuel injection; battery costs are still projected to be at least three times target costs; projected costs for fuel-cell systems are still at least five times the long-term targets; low-cost manufacturing techniques have yet to be developed for power electronics; and integrated thermal management for power electronics is still complex and costly. As far as the committee is aware, detailed cost analyses for overall vehicle systems have not even been attempted yet by some of the automotive companies, but rudimentary estimates for complete vehicle systems show cost penalties of several thousand dollars. In fact, DaimlerChrysler has estimated a $7,500 selling price cost penalty for a production version of its ESX3 concept vehicle.

As the committee has stated in previous reviews, the cost penalty is exacerbated with an HEV, which is more complex than a nonhybrid vehicle. If the federal administration and Congress want to promote the deployment of high fuel economy PNGV-type vehicles, they may have to evaluate the advisability of providing temporary incentives (e.g., tax rebates) to offset higher initial vehicle costs.

Impact of Emission Standards

The Tier 2 NO_x and PM emission standards announced at the end of 1999 are significantly more stringent than the standards in place when the PNGV program was initiated. Based on several presentations, including two by EPA, the feasibility of meeting these standards with acceptable fuel economy in the time frame of the PNGV program is very questionable. Meeting the Tier 2 standard for diesel engines will likely require new catalytic materials and new emissions control concepts. In the committee's judgment, the Tier 2 standards as currently promulgated could potentially preclude the early introduction of the CIDI internal combustion engine with its significant fuel economy benefit in the United States or, as a result of compromises to meet the Tier 2 standards, reduce the CIDI engine fuel economy benefit. To meet new standards, PNGV may have to shift its attention from the CIDI engine toward the adaptation of other internal combustion engines with better potential for extremely low emissions but at lower energy conversion efficiencies and with higher carbon dioxide emissions. The change in emission standards will affect the PNGV program in a variety of ways, which are discussed throughout the report.

The introduction of Tier 2 standards raises questions that should be discussed by USCAR and government agencies involved in the PNGV, namely whether the reduced fuel economy that would be provided by the alternative engines available for the next phase of the program (i.e., the 4SDI [four-stroke direct-injection] spark-ignition engine and the port fuel-injected gasoline spark-ignition engine), which have significantly better potential for meeting the Tier 2 emission levels, is an appropriate trade-off from a national perspective. A wiser decision might be to extend the deadline for meeting these two objectives—improved fuel economy and lower emissions—allowing more time for the development of new fuel economy technology. This issue will have to be resolved before PNGV can clarify the objectives of the production-prototype phase of the program.

Fuel Issues

Although modern in-cylinder injection, high rail pressures, and closed-loop control can dramatically reduce emissions from CIDI engines, the new Tier 2 standards are not likely to be met with currently available high-sulfur diesel fuel, even with PM traps and NO_x absorbers. For fuel-cell vehicles, onboard reforming of gasoline to produce hydrogen would probably require substantial changes in gasoline composition; in fact, optimum performance would probably require a new hydrocarbon fuel very different from conventional gasoline. In addition, because of the negative impact of sulfur on catalysts, sulfur concentrations in all petroleum fuels would have to be reduced substantially. As the committee has pointed out in previous reports, significant changes in fuels would also have wide-ranging effects on the petroleum and fuels industry, infrastructure, and costs. Therefore, changes in fuels must be carefully planned and their implications thoroughly investigated.

The PNGV program has been devoting more attention to fuel composition issues and is working with individual petroleum companies but has not established a mechanism to determine the commercial trade-offs between engine systems (with the attendant effects on fuel economy) and fuel compositions. Greater participation by the petroleum and fuels industries will be critical to the success of the PNGV program.

Up to now, PNGV has given a high priority to automotive systems that can achieve the goals of the program. The committee agrees that this was the correct approach for the beginning stages of the program. Now, however, fuel issues must be addressed strategically, in cooperation with the petroleum industry. Otherwise, commercialization of the technologies being developed could be delayed because of the lead time required to manufacture and distribute modified fuels.

FUEL CELLS

Since the beginning of the PNGV program, the fuel-cell energy converter has shown the potential to provide high fuel economy and produce very low emissions. Despite the significant progress that has been made and the substantial private sector resources that have been expended, the committee continues to consider fuel cells a long-range technology applicable to automobiles beyond 2004 because of the outstanding technical and cost issues. However, because the Tier 2 standards have increased the development risk for the CIDI engine, the fuel cell has been elevated to a higher level of importance, and PNGV has been focusing its efforts on processing (reforming) gasoline on board the vehicle to produce hydrogen, which would sidestep the fuels infrastructure problem of the lack of widespread availability of hydrogen. However, an onboard fuel processor would complicate the vehicle system immensely and reduce vehicle fuel economy; storing hydrogen onboard the vehicle would simplify the vehicle system somewhat, but a high-density storage system would have to be developed, as well as an infrastructure, to ensure that hydrogen would be widely available at service stations. All of these trade-offs will have to be analyzed as the program moves forward, and several technology options should be pursued as the basis for a technically, economically, and logistically feasible combination of fuel infrastructure and fuel cell energy converter.

ADEQUACY AND BALANCE OF THE PNGV PROGRAM

Government-sponsored R&D on advanced automotive technologies is primarily being done by the U.S. Department of Energy (DOE). DOE is expected to provide about $128 million in fiscal year 2000 for PNGV; the program also receives funding from other sources. Consequently, DOE's funding allocation among technologies, therefore, does not represent the distribution of effort among technologies by PNGV as a whole. Other government agencies (e.g., U.S. Department of Commerce, EPA, the U.S. Department of Transportation, and the National Science Foundation) are expected to provide approximately $110 million of relevant funding, about half of which will be used for emission control projects and half for long-range research.

The committee was informed by PNGV that USCAR had solicited, on a confidential basis, overall figures on expenditures by its members for "PNGV-related" research. These figures show a total investment by the three companies for 1999 of $982 million. Estimates of investments for each of the previous three years was comparable. This very large investment (far higher than the program's expected 50/50 government/industry matching level) represents major efforts on the part of the industry partners to develop the year 2000 concept cars.

The adequacy of funding for PNGV is difficult to assess because the data provided to the committee are incomplete. Because progress toward meeting Goal 3 appears insufficient to meet the objectives by 2004, one might conclude that resources are inadequate. However, the USCAR partners indicated to the committee that the lack of talented people is a greater handicap than the lack of adequate funding and that they need new ideas (breakthroughs) more than dollars. The committee is inclined to agree because, although increased funding might speed up some projects and also support a broader program more likely to achieve breakthroughs, no important areas of the program seem to be starving for funds. Thus, at the moment, progress in the PNGV program will depend more on inventive solutions and the influx of additional technical people than on increases in resources.

The balance of resources in PNGV is even harder for the committee to assess than funding because no data are available on industry distribution of funding by project. The balance represented in the government programs appears to be appropriately weighted to long-range R&D. The committee assumes that during the development of the concept vehicles, industry resources have been directed primarily toward solving the problems of hybrid vehicle optimization. However, from 2000 to 2004, a different balance will be required, depending on which course individual companies choose to follow. For example, if a company chooses to continue working toward the development of a production prototype vehicle with a CIDI engine in an HEV configuration, then a major effort must be mounted on emissions control of that power plant and a determination made of the benefits of optimizing that system for emissions control rather than for efficiency (i.e., fuel economy and cost penalties involved in meeting the Tier 2 emission standards). If a company chooses to replace the diesel engine with a gasoline spark-ignition engine, which the USCAR partners have indicated can meet the Tier 2 standards, then the optimized fuel economy of that configuration and the diesel system could be compared. The gasoline system would probably have a somewhat smaller cost penalty than the diesel configuration. The committee believes that PNGV should investigate both of these options, using the best systems analysis and experimental evidence available. The balance of future R&D should then be adjusted according to the results.

ISSUES FOR PNGV BEYOND 2000

The committee encourages PNGV leadership to develop specific objectives for the production-prototype phase of the program with the following goals in mind. First, each automotive company member should develop production-feasible total vehicle concepts that come as close as is practical to the original vehicle performance objectives of Goal 3 (i.e., meeting the mandated emission requirements, balancing the inevitable shortfalls in fuel economy, vehicle performance, and affordability to maximize potential market acceptability). Second, the automotive partners should develop production-feasible versions of new PNGV component technologies that can, in a more evolutionary way, be incorporated into new vehicle designs under Goal 2. The first objective would continue to "stretch" the new technologies and system concepts that, together, have the potential to introduce substantially better fuel economy into the vehicle fleet. The second objective would encourage the development and application of component technologies critical to improving fuel economy by encouraging their early commercial introduction.

SELECTED RECOMMENDATIONS

Recommendation. PNGV should quantify the trade-off between efficiency and emissions for the power plants under consideration. The PNGV systems-analysis team should attempt to develop and validate vehicle emissions models of sufficient sophistication to provide useful predictions of the emissions potential for a variety of engines (e.g., the compression-ignition direct-injection engine, the gasoline direct-injection engine) and exhaust gas after-treatment systems in various hybrid electric vehicle configurations. The models could be used to help PNGV evaluate the feasibility of meeting the Environmental Protection Agency's Tier 2 emissions standards with various vehicle system configurations. These data should then be used to establish an appropriate plan for the next phase of the program.

Recommendation. At this stage, PNGV should direct its program toward an appropriate compromise between fuel economy and cost using the best available technology to ensure that a market-acceptable production-prototype vehicle can be achieved by 2004 that meets Tier 2 emission standards.

Recommendation. Given the potential of fuel-cell technology for meeting the efficiency and emissions objectives of the PNGV program, the systems-analysis team should increase its efforts to develop more complete and accurate fuel-cell system and component models to support the development and assessment of fuel-cell technology.

Recommendation. In the area of fuel cell development, PNGV, and especially the U.S. Department of Energy, should emphasize high-risk, high-payoff research in critical areas, such as fuel processing, carbon monoxide-tolerant electrodes, and air-management systems.

Recommendation. PNGV should continue to work on cell chemistry of lithium battery systems to extend life and improve safety, while continuing to lower costs. Performance and cost targets should be refined as overall vehicle systems analysis determines the optimal degree of vehicle hybridization.

Recommendation. As the PNGV program moves toward the 2004 production-prototype milestone, affordability will be a key requirement. Therefore, the development of an efficiently designed and fabricated steel-intensive vehicle being worked on by the American Iron and Steel Institute in the Ultralight Steel Autobody-Advanced Vehicle Concepts (ULSAB-AVC) project should be closely followed, and the possibility of applying the ULSAB concepts to a hybrid steel-aluminum vehicle should be explored.

Recommendation. The committee recognizes the cost reduction potential of DaimlerChrysler's thermoplastic composite injection-molding technology and urges that this work be continued to bring the technology to successful commercialization.

Recommendation. The committee regards structural crashworthiness, and safety in general, in the design of lightweight PNGV vehicles as extremely important. Using the Oak Ridge National Laboratory car-to-car collision simulation capability, the National Highway Traffic and Safety Administration should support a major study to determine how well lightweight PNGV vehicles would fare in collisions with heavier vehicles and to assess potential improvements.

Recommendation. Defining automotive system/fuels trade-offs and establishing a basis for ensuring that the required fuels are available as higher efficiency vehicles become commercially available will require extensive cooperation among automotive and petroleum industry representatives at all levels of responsibility. Therefore, PNGV should strengthen and expand its cooperative efforts with the petroleum industry, including issues related to fuels for fuel cells. Government leadership will be necessary to initiate this cooperative effort and provide incentives for petroleum company involvement.

Recommendation. PNGV should consider conducting a comprehensive assessment of the consequences of fuel choices for fuel cells and their impact on PNGV's direction and ultimate goals. PNGV should assess the opportunities and costs for generating hydrogen for fuel cells at existing service stations and storing it on board vehicles and compare the feasibility, efficiency, and safety of this option with onboard fuel reforming. This study would help PNGV determine how much additional effort should be devoted to the development of onboard fuel reforming technologies.

1

Introduction

On September 29, 1993, President Clinton initiated the Partnership for a New Generation of Vehicles (PNGV) program, a cooperative research and development (R&D) program between the federal government and the United States Council for Automotive Research (USCAR), whose members are DaimlerChrysler Corporation, Ford Motor Company, and General Motors Corporation (GM).[1] The purpose of the PNGV program is to improve substantially the fuel efficiency of today's automobiles and enhance the U.S. domestic automobile industry's productivity and competitiveness. The objective of the PNGV program is to develop technologies for a new generation of vehicles with fuel economies up to three times (80 miles per equivalent gallon of gasoline [mpg]) those of a 1994 comparable midsize sedan, comparable performance, size, utility, and cost of ownership and operation (adjusted for economic changes), and able to meet or exceed federal safety and emissions requirements (The White House, 1993).

The PNGV declaration of intent includes a requirement for an independent peer review "to comment on the technologies selected for research and progress made." In response to a written request by the Undersecretary for Technology Administration, U.S. Department of Commerce, on behalf of PNGV, in July 1994 the National Research Council established the Standing Committee to Review the Research Program of the Partnership for a New Generation of Vehicles. The committee conducts annual independent reviews of PNGV's research program,

---

[1] USCAR, which predated the formation of PNGV, was established to support intercompany, precompetitive cooperation to reduce the cost of redundant R&D in the face of international competition. Chrysler Corporation merged with Daimler Benz in 1998 to form DaimlerChrysler. USCAR is currently comprised of a number of consortia as shown in Appendix D.

advises government and industry participants on the program's progress, and identifies significant barriers to success. This is the sixth review by the committee; the previous studies are documented in five National Research Council reports, which also contain background on the PNGV program and the committee's activities (NRC, 1994, 1996, 1997, 1998, 1999).

The PNGV goals, and the basis for all of the National Research Council reviews, are articulated in the PNGV Program Plan (PNGV, 1995):

> **Goal 1. Significantly improve national competitiveness in manufacturing for future generations of vehicles.** Improve the productivity of the U.S. manufacturing base by significantly upgrading U.S. manufacturing technology, including the adoption of agile and flexible manufacturing and reduction of costs and lead times, while reducing the environmental impact and improving quality.
>
> **Goal 2. Implement commercially viable innovations from ongoing research on conventional vehicles.** Pursue technology advances that can lead to improvements in fuel efficiency and reductions in the emissions of standard vehicle designs, while pursuing advances to maintain safety performance. Research will focus on technologies that reduce the demand for energy from the engine and drivetrain. Throughout the research program, the industry has pledged to apply those commercially viable technologies resulting from this research that would be expected to increase significantly vehicle fuel efficiency and improve emissions.
>
> **Goal 3. Develop vehicles to achieve up to three times the fuel efficiency of comparable 1994 family sedans.** Increase vehicle fuel efficiency to up to three times that of the average 1994 Concorde/Taurus/Lumina automobiles with equivalent cost of ownership adjusted for economics.

As the committee has noted in previous reports, and as has been noted in a number of other studies, achieving significant improvements in automotive fuel economy and developing competitive advanced automotive technologies and vehicles can provide important economic benefits to the nation, improve air quality, improve the nation's balance of payments, and reduce emissions of greenhouse gases to the atmosphere (DOE, 1997; NRC, 1992, 1997, 1998; OTA, 1995; PCAST, 1997; Sissine, 1996). Although U.S. gasoline prices have risen in recent months, they are relatively low in real terms, and U.S. automobile purchasers have little incentive to consider fuel economy as a major factor in their purchase decisions. In addition, the sales of light trucks, especially sport utility vehicles, which have lower legislated fuel economy requirements than automobiles, continue to increase. The lack of market incentives in the United States for buyers to purchase vehicles with high fuel economy has made it difficult to realize public benefits from improvements in fuel economy.

The PNGV strategy of developing an automobile with a fuel economy of up to 80 mpg that maintains current performance, size, utility, and cost levels while meeting safety and emissions standards would circumvent the lack of economic incentives for buying automobiles with high fuel economy. If the PNGV strategy is successful, buyers will purchase vehicles with all of the desirable consumer attributes, as well as greatly enhanced fuel economy. The development of this vehicle, as the committee has noted in previous reports, is extremely challenging. But this ambitious goal and the PNGV program have stimulated the rapid development worldwide of the required technologies, which highlights the potential strategic value of programs like PNGV and the importance of staying abreast of developments in foreign technology. If the Goal 3 vehicle does not quite achieve the triple-level fuel economy but approaches the cost and performance objectives, it should still be far more fuel efficient than current vehicles, which would be an outstanding achievement.

PNGV's objective is to bring together the extensive R&D resources of the federal establishment, including the national laboratories and university-based research institutions, and the vehicle design, manufacturing, and marketing capabilities of both the USCAR partners and suppliers to the automotive industry.[2] In general, government funding for PNGV is primarily used for the development of long-term, high-risk technologies. Funding by USCAR and industry is generally used for the development of technologies with nearer term commercial potential, the implementation of government technology developments into automotive applications, and the production of concept vehicles. Substantial in-house proprietary R&D programs are also under way at USCAR partners' facilities.

Technical teams responsible for R&D on the candidate subsystems, such as fuel cells, four-stroke direct-injection (4SDI) engines, and others, are central to PNGV. A manufacturing team, an electrical and electronics power-conversion devices team, a materials and structures team, and a systems-analysis team are also part of the PNGV organization (NRC, 1996, 1997, 1998, 1999). Technical oversight and coordination are the responsibilities of a vehicle-engineering team, which provides the technical teams with vehicle-system requirements supported by the systems-analysis team.

PROGRAM MILESTONES

At the inception of the program, several milestones were established: a technology selection in 1997; concept vehicles in 2000; and production prototypes in 2004. At the end of 1997, PNGV reached a critical milestone of technology selection (often referred to as the technology "downselect" process) based on assessments of system configurations for alternative vehicles. In this process, several technology options, such as gas turbines, Stirling engines, ultracapacitors for energy storage, and flywheels for energy storage, were eliminated as leading

---

[2] The U.S. Department of Commerce, the U.S. Department of Energy (DOE), the U.S. Department of Transportation, the Environmental Protection Agency (EPA), and the U.S. Department of Defense are the federal agencies involved in the PNGV program.

candidates. In its fourth review, the committee agreed with PNGV's technology selections (e.g., four-stroke, internal-combustion engines, fuel cells, batteries, power electronics, and structural materials). The four-stroke compression ignition direct injection (CIDI or diesel) engine is the primary power plant with the potential to meet the fuel economy goals within the time frame of the program; the fuel cell power plant is considered a longer range technology. After the technology selection process, PNGV was able to concentrate its resources on fewer technologies with the intent of defining, developing, and constructing concept vehicles by 2000 and production prototypes by 2004 (PNGV, 1995). Using PNGV-developed technologies and their own in-house proprietary technologies, the USCAR companies each developed separate concept vehicles (see Chapter 4). Thus, the program as a whole will not design or build a joint concept car, a decision that the committee supports.

Although the technologies most likely to result in concept and production prototype vehicles that could meet the Goal 3 requirements were selected during the downselect process, as other, longer range technologies continue to evolve, they may be incorporated into subsequent concept vehicles, as appropriate, and a series of concept vehicles will probably be developed after 2000. Since the beginning of the program, PNGV has addressed many technology areas, including advanced lightweight materials and structures; efficient energy-conversion systems (including advanced internal combustion engines, gas turbines, Stirling engines, and fuel cells); hybrid electric propulsion systems; energy-storage devices (including high-power batteries, flywheels, and ultracapacitors); emission control systems; efficient electrical and electronic systems; and systems to recover and utilize exhaust energy and braking energy.

SCOPE OF REVIEW

The 15-member committee that conducted this sixth review of PNGV had a wide variety of expertise (see Appendix A for biographical information). The committee was asked to address the following tasks in this review:

(1) In light of the PNGV program technical goals and previous NRC Standing Committee recommendations, examine and comment on the overall balance and adequacy of the PNGV research effort, the rate of progress, and the technical objectives and schedules for each of the major technology areas (i.e., fuel cells, 4-stroke direct injection engines and emissions control, power electronics and electrical systems [electric drive], energy storage, and structural materials).

(2) In light of the proposed emission requirements for the 2004–2010 time period, examine and comment on the ongoing fuels, propulsion engine, and emission control research efforts to identify and develop commercially viable very low emission (e.g., California LEV-2 and anticipated EPA Tier 2 standards) propulsion systems based on fuel cells and 4-stroke direct injection engines.

(3) If invited by DaimlerChrysler Corporation, Ford Motor Company, or General Motors Corporation, committee subgroups will review each of the company's technology application, vehicle integration, and supporting advanced manufacturing technology efforts directed at the year 2000 PNGV concept vehicles.

The conclusions and recommendations in this report are based on the committee's meetings, presentations, and other data-gathering activities (see Appendix C). Some of the material reviewed by the committee was presented by USCAR as proprietary information under an agreement signed by the National Academy of Sciences, USCAR, and the U.S. Department of Commerce (on behalf of the federal government).

Opinions of the goals of any program are bound to differ. Some think the PNGV time frame should be longer to allow for the maturation of some longer range technologies; some think the goals should be to reduce the combined emissions of greenhouse gases rather than fuel economy. As the committee has noted in previous reports, these reviews were based on the vision, goals, and target dates for PNGV articulated by the president and the R&D programs that have been launched (NRC, 1996, 1997, 1998, 1999). Assuming that PNGV partners will seriously pursue the objectives of the program, the committee understands its role as providing independent advice to help PNGV achieve its goals. Therefore, the committee has tried to identify actions that could enhance the program's chances for success and has refrained from making judgments on the value of the program to the nation. The goals were accepted as given, including goals 1 and 2, which, unlike Goal 3, are open-ended and do not have quantitative targets and milestones. The objectives of goals 1 and 2, in many instances, support progress toward Goal 3, especially the development of manufacturing capabilities for the advanced automotive technologies being considered for the Goal 3 vehicle. The goals are summarized in the updated PNGV Technical Roadmap (PNGV, 1999a).

Because regulations and the market continue to change, however, the committee believes that PNGV should periodically reassess its objectives, especially in light of the negative impact the new emissions standards (i.e., the Tier 2 emissions standards) may have on the success of the program. The final regulations for Tier 2 standards announced by the Environmental Protection Agency (EPA) on December 21, 1999, greatly increase the development challenge and risk for the CIDI engine, the primary power plant now under consideration, to meet the fuel economy goal of 80 mpg in PNGV's time frame. Alternative power plants that could meet the PNGV 2004 time frame would probably result in vehicles with a lower fuel economy (see Chapter 5 for further discussion).

2

Development of Vehicle Subsystems

CANDIDATE SYSTEMS

The success of the PNGV program will depend on the integration of R&D programs that can collectively improve the fuel efficiency of automobiles and meet the requirements of comparable size, reliability, durability, safety, and affordability of today's vehicles. At the same time, the PNGV vehicles must meet even more stringent emission standards and recycling levels and must use components that can be mass produced and maintained in a manner similar to current automotive products.

In order to achieve the Goal 3 fuel economy objective of 80 mpg, the energy conversion efficiency of the chemical conversion system (e.g., a power plant, such as a CIDI engine or a fuel cell) averaged over a driving cycle will have to be at least 40 percent, approximately double today's efficiency. In addition to improving the energy-conversion efficiency of the power train (including energy converters and transmissions) and reducing other energy losses in the vehicle, achieving the Goal 3 fuel economy objective will require a very large weight reduction (> 40 percent) through new vehicle designs and lightweight materials.

In short, meeting the Goal 3 fuel economy target will require extensive innovation. For example, the primary power plant, when used in a hybrid electric vehicle (HEV) configuration, will have to be integrated with energy-recovery and energy-storage devices. Radically new vehicle structures will be necessary to reduce vehicle weight. The design and performance of essentially every aspect and function of the vehicle, from the passenger heating and cooling systems to the conversion efficiency of the exhaust-gas after-treatment systems, will have to be evaluated for possible energy savings.

Hybrid drive systems, the PNGV near-term power trains of choice, use energy-storage devices to reduce fluctuations in demand on the primary power plant, thereby permitting the use of a smaller power plant operating closer to optimum conditions for increased energy-conversion efficiency and reduced emissions. These storage devices also allow recovery of a portion of the vehicle's kinetic energy during braking operations.

In this chapter, the committee continues its evaluation of the candidate energy-conversion and energy-storage technologies that survived the 1997 technology selection process, as well as of candidate electrical and electronic systems and advanced structural materials for the vehicle body.

The committee reviewed R&D programs for: four-stroke internal-combustion reciprocating engines, fuel cells, electrochemical storage systems (rechargeable batteries), power electronics and electrical systems, and structural materials and safety to assess their progress toward commercial applicability. In the committee's opinion, PNGV has continued to make significant progress on the development of candidate systems and the identification of critical technologies that must be addressed to make each system commercially viable.

Indeed, global competitiveness is one of the objectives of the PNGV program. In fact, previous committee reports have specifically addressed the directions and trends of international development programs. The committee is aware that the PNGV partnership continues to maintain awareness of international programs and, in many cases, the USCAR partners are participating in those developments through international operations and coalitions. Therefore, a separate section on international developments has not been included in this report.

INTERNAL COMBUSTION RECIPROCATING ENGINES

The internal combustion engine continues to be the primary candidate power plant for meeting near-term PNGV program goals. To meet the fuel economy target of Goal 3, the internal combustion engine will have to be integrated into an HEV configuration. The CIDI or diesel engine is the most fuel efficient internal combustion engine being developed. Consequently, in the near term, maximizing the efficiency of a diesel engine and integrating it into an HEV will provide maximum vehicle efficiency. However, as will be discussed below, the challenges of meeting the new California Air Resources Board (CARB) and EPA Tier 2 emission standards have called into question the viability of the CIDI engine as the primary energy converter in PNGV's time frame. As a result of the new standards, the technical team working on the 4SDI (four-stroke direct-injection) engine began focusing more on emissions control research for the CIDI engine. In addition, they are continuing research programs on other combustion systems for internal combustion engines, namely the homogeneous-charge compression-ignition engine, the gasoline direct-injection engine, and the homogeneous-charge spark-ignition engine.

Figure 2-1, which was included in previous committee reports, is a convenient summary of the advantages and challenges of the internal combustion engines being considered as primary energy converters for the PNGV program.

Program Status and Plans

In the past year, the 4SDI technical team has made excellent progress in continuing the development of a power system to meet the PNGV goals. In addition, each USCAR partner developed and built its own concept car. Based on proprietary visits to each partner's laboratories to be briefed on concept-car programs, the committee recognizes and commends the tremendous and impressive efforts by each USCAR partner to develop and build these concept cars and integrate promising technologies into the vehicles. The extent to which each concept vehicle represents a different approach toward meeting the PNGV goals is testimony to the dedication, creativity, and industry of the USCAR partners.

All three partners are developing diesel-powered electric hybrid systems for their concept vehicles. However, overall vehicle objectives and the logical basis for hybridization vary among the partners, whose approaches to trade-offs among fuel economy, degree of hybridization, emission reduction strategies, attaining manufacturability, and cost reduction are quite different. Because of the importance of these vehicles as a program milestone, a separate section of this report is devoted to them (see Chapter 4).

In addition to proprietary internal work on the concept vehicles, PNGV continued its collaborative programs on precompetitive fundamental technologies, including interactions between fuel composition and engine performance (as recommended in the committee's fifth review [NRC, 1999]); the fundamentals of combustion, including some very novel concepts; and in response to Tier 2 emission standards, a stepped-up investigation of emission control technologies and sensors. Progress to date and the future plans of the 4SDI technical team are discussed below in three sections: engine-fuel interactions; engine-combustion system developments; and after-treatment and controls and sensors.

Engine-Fuel Interactions

The effects of fuel chemistry and physical properties on engine performance and emissions was the focus of intense activity during the past year motivated by data reported in the literature showing that fuel composition affects nitrogen oxides (NO_x) and particulate matter (PM) emissions from a variety of diesel engines (Ryan et al., 1998; Takatori et al., 1998; Tanaka and Takizawa, 1998; Wall and Hoekman, 1984). The data also indicated that fuel composition could affect the performance of exhaust-gas after-treatment systems. However, the extent to which these changes would carry over to a state-of-the-art, highly

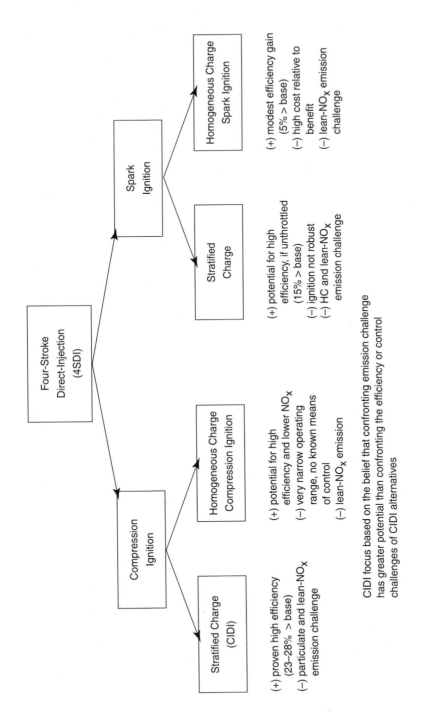

FIGURE 2-1 Alternatives for energy conversion. The baseline engine is the port fuel injection homogeneous charge spark injection for model year 1993. Source: Low et al., 1999.

controllable engine system in which the engine and fuel have been optimized together was not clear. Similarly, as new exhaust-gas after-treatment systems are developed, it will be important to determine whether they retain their sensitivity to fuel composition. These are the general areas of focus for PNGV's studies on fuel-engine systems and exhaust-gas after-treatment systems.

PNGV is working cooperatively with the low-emission partnership of USCAR, the fuels industry, the U.S. Department of Energy (DOE), and the national laboratories to evaluate the effects of fuel composition and physical properties on engine performance and exhaust-gas after-treatment conversion efficiencies. The working relationships range from formal (e.g., the Engine Manufacturers Association-EPA-American Petroleum Institute Diesel Fuel for 2004 Program and the DOE Advanced Fuels Technology Initiative) to ad hoc (e.g., arrangements between individual USCAR partners with fuel companies [BP Amoco, Shell, and ExxonMobil] to include fuel composition in their matrix for in-house engine development programs).

Collaborative in-house testing by PNGV teams and their energy company partners is being done on a CARB commercial #2 diesel; a petroleum-based, low-sulfur (<10 ppm), low-aromatic (<9 percent by weight) fuel; a neat Fischer-Tropsch diesel fuel; and an oxygenated fuel made by blending 15 percent dimethoxymethane with the low-sulfur, low-aromatic petroleum-based fuel. Standard fuel properties, such as distillation, cloud point, pour point, density, and viscosity were measured for all of the fuels tested. The choice of fuels was based on practical considerations of what might be feasible to introduce into the current fuel infrastructure.

The fuels were run in four different state-of-the-art engines by DaimlerChrysler, Ford, and GM at a range of speeds and loads. At the time of this review, the initial testing had been completed and the data were being analyzed. Although the detailed results and conclusions have not been released yet, one conclusion seems apparent. No clearly identifiable practical fuel composition or property will enable the CIDI engine to meet the new emission standards simply through improved combustion. Although control of the fuel composition and/or properties yields some benefits, the gains in a highly controlled and optimized engine were not as large as expected.

The effects of fuel properties on exhaust-gas after-treatment systems are also being investigated. As the committee has noted in previous reports, the PNGV technical team has determined that none of the 4SDI technologies will be capable of meeting the emission standards without extensive exhaust-gas after-treatment (NRC, 1998, 1999). Therefore, determining the effects of fuel properties, especially sulfur content, on the conversion efficiency of the after-treatment devices will be critical.

PNGV's activities in the last year in this area have been intense. PNGV teams are working cooperatively with the Diesel Emission Control Sulfur Effects

(DECSE) test program, which is attempting to determine the impact of diesel fuel sulfur level on emission control systems. The DECSE program is evaluating the effects of fuel sulfur concentrations ranging from 3 to 350 ppm on NO_x absorbers, PM traps, lean-NO_x systems, and diesel oxidation catalysts. The data are still being analyzed, but a consensus has already emerged that sulfur degrades the performance of exhaust-gas after-treatment systems.

This consensus is supported by the results of fundamental research on advanced after-treatment technologies at the national laboratories (discussed in more detail below). When researchers incorporate sulfur concentration into their test matrices, the data show a strong correlation between the system conversion efficiencies and sulfur concentration. In general, the less sulfur the better. But, because the cost-benefit analysis of incremental sulfur removal has not been conclusively evaluated, work in this area should be continued.

PNGV plans to continue its very active program in the area of fuel-engine system interactions. Ad hoc working relationships with fuel companies, as well as participation in formal programs, are critical aspects of the program on fuel-engine system interactions. These ad hoc relationships seem to be working well and provide flexibility to the individual companies to pursue appropriate research.

Engine-Combustion System Developments

Understanding the challenges facing PNGV requires understanding the fuel economy and emission characteristics of different engine types. To date, the approach of the 4SDI technical team has been to pursue technologies that promise maximum fuel efficiency and then address the more stringent emission targets. For an internal combustion engine, the maximum work per unit fuel is obtained when the engine is operated in a fuel-lean (i.e., excess air) condition. If the engine load can be controlled by varying the quantity of fuel introduced into the cylinder without altering the amount of inducted air, energy losses associated with drawing air into the engine across a throttle (throttling losses) can be eliminated. Throttling losses in a typical homogeneous spark-ignition engine range from a few percent at high loads to more than 90 percent at very light loads.

The desirable engine attributes described above are inherent characteristics of diesel engines, the most efficient internal combustion engines available. For this reason, as well as because extensive manufacturing and operating experience are available, the diesel engine is the engine of choice as the principal energy converter for the PNGV program. However, because the diesel engine attains lean combustion by maintaining a heterogeneous combustion system, there are large gradients in air-fuel mixtures in the combustion chamber during energy release. As a result, PM forms in fuel-rich regions, and NO_x forms in high-temperature stoichiometric regions. The challenge is to control NO_x and PM emissions from diesel engines simultaneously.

The implementation of the EPA Tier 2 standards and the necessity of meeting CARB Ultra Low Emission Vehicle (ULEV) II requirements in California, as well as in other states that may choose to adopt them, has significantly increased the challenge of meeting emission regulations with any internal combustion engine, especially a diesel engine. The initial design targets for NO_x and PM emissions at the start of the PNGV program were 0.2 g/mile NO_x and 0.04 g/mile PM. These targets seemed to be attainable with an advanced technology diesel engine and some exhaust-gas after-treatment. In October 1997, immediately prior to the PNGV technology selection process (downselect), new research targets were introduced, namely 0.20 g/mile NO_x and 0.01 g/mile PM. These levels required substantial improvements in the efficiency of combustion and after-treatment systems. At that time PNGV introduced R&D on the effects of fuel composition into the program.

The new EPA Tier 2 standards tighten the emission requirements further. The new standards mandate fleet averages of 0.07 g/mile NO_x and 0.01 g/mile PM, including light trucks and sport utility vehicles (SUVs) of up to 8,500 pounds. The Tier 2 regulations set a phase-in time schedule for different classes of vehicles, as well as emission level "bins" into which different vehicles can be grouped, as long as the fleet average emission level is met. This regulatory flexibility is intended to allow for emerging technologies being introduced into the market by giving manufactures time to gain production experience and continue their development. However, for each vehicle that exceeds the fleet average emission level, a number of vehicles with emission levels below the fleet average will have to be sold to maintain the fleet average. Producing ultralow emission vehicles will be a substantial challenge, and, from a business perspective, if the manufacturer's lightest and most fuel-efficient vehicle, the PNGV vehicle, cannot meet the mandated fleet average, in all likelihood it will not be pursued. Clearly, PNGV needs to reassess the technological avenues and roadblocks to attaining fleet average standards with the diesel-powered HEV and decide whether or not it should be the primary choice for the PNGV vehicle.

In an attempt to put current work on the diesel-powered HEV in perspective, the 4SDI team assessed to what extent the current projections of after-treatment system performance would have to be modified to meet the new emission standards. The results of this sobering exercise are summarized below:

- With a gasoline port-fuel-injected, stoichiometric, homogeneous-charge engine, the emission standards could probably be met with advanced three-way catalysts of different combinations.
- With a direct-injection, spark-ignited gasoline engine as the principal energy converter, advanced NO_x traps and PM traps would be required.
- With a diesel engine, exhaust-gas after-treatment systems would require 75 percent and 50 percent reduction efficiencies for NO_x and PM, respectively.

- Because the fuel economy projections are based on extrapolations of emissions reduction technology, the fuel economy penalty associated with meeting the emission standards could not be quantified.[1]

The immediate effect of the announcement of the Tier 2 standards has been a shift in resources toward after-treatment technologies to reduce emissions and away from improvements in fuel economy. Prior to the announcement, PNGV believed that the best chance of reaching 80 mpg in its time frame was with a diesel engine as the primary energy converter. Based on the current assessment, the challenges of meeting the Tier 2 emission standards with the diesel engine will be enormous, and the technology most likely to meet the new standards is an engine that uses homogeneous stoichiometric combustion, for which the exhaust-gas after-treatment is most advanced. Unfortunately, this type of engine is the least efficient of all of the candidate engines being investigated. Ironically, in the near term, the likely effect of the Tier 2 emission standards will be to promote the development of the less efficient engine systems.

Despite the increased emphasis on emission reduction technologies, which are being pursued primarily through improved exhaust-gas after-treatment systems, the 4SDI technical team has also continued to perfect in-cylinder combustion processes. This R&D is being performed by the PNGV partners in collaboration with DOE national laboratories, the Department of Defense Tank Automotive Command, EPA, AVL List GmbH, FEV, Wayne State University, and the University of Wisconsin at Madison. Projects have been divided into two categories: (1) enabling technologies that will be critical in the near term (less than three years); and (2) technologies that could be important in the longer term (more than three years).

Near-term projects include: the development of advanced fuel-injection systems, studies of the cylinder-to-cylinder distribution and transient response of exhaust-gas recirculation (EGR); the development of pressure-reactive pistons for dynamic changes in the compression ratio; and detailed comparisons of combustion data obtained in an optical research engine, a similar metal engine, and the predictions of sophisticated, three-dimensional computer simulations. The results of these projects during the past year have varied.

To date, using injection-rate shaping and multiple injection events for emissions control have shown no significant reduction in emissions. However, this

---

[1] For example, the diesel engine operates in an air-fuel regime that is overall fuel lean (i.e., with excess air). The combustion process is not premixed or homogeneous. Injecting the fuel directly into the combustion chamber results in large gradients of air-fuel mixture. The cylinder contents vary spatially from extremely lean through stoichiometric to very rich air-fuel mixtures. In addition, the fuel economy penalty of the NO_x trap or lean-NO_x catalytic reduction technology has not been quantified.

technology has not been completely explored. In addition, this technology may have a critical effect on the composition of the exhaust gas that enters the exhaust-gas after-treatment system. A special fuel system using dimethyl ether was successfully constructed and tested, and more extensive testing of this interesting fuel can now be done in a Ford DIATA (direct-injection, aluminum-block, through-bolt assembly) diesel engine.

An instrumentation system for monitoring the transient EGR distribution was developed by FEV this year, which will enable the detailed investigation of the cylinder-to-cylinder distribution of EGR and the transient response of the engine-EGR system. Because EGR is an important method of in-cylinder NO_x control, the results of this investigation will be significant.

Through the collaborative efforts of Sandia National Laboratories (SNL), Wayne State University, and the University of Wisconsin at Madison, the program on the optical engine, the metal engine, and advanced simulation is now fully operational. Data showing soot luminosity and liquid fuel distribution were mapped in the optical engine, and tests of EGR tolerance, from zero to the maximum possible amount, were performed in the metal engine. Both of these data sets were used as a basis of comparison for the predictions of the simulation. The objective of the project is to characterize fully the in-cylinder flow field with the effects of swirl and to validate the full combustion simulation.

Projects in the longer time frame include: continued investigation of homogeneous-charge, compression-ignition combustion; studies on advanced variable compression-ratio engines; studies of electromagnetically actuated engine valves (the camless engine); and the investigation of several novel engine concepts. Progress was also made in these projects.

Researchers at the Combustion Research Facility at SNL in Livermore, California, are close to completing a dedicated laboratory for studying the fundamentals of homogeneous-charge, compression-ignition combustion. Some preliminary data have already been obtained. Experiments and comparisons of the results with predictions from detailed chemical-kinetic computer codes are planned for this facility.

PNGV also continues to investigate novel engine concepts, such as a variable compression-ratio small engine that would operate either as a port fuel-injected or gasoline direct-injected engine. The idea behind this concept is to minimize throttling by varying the compression ratio and boost pressure. If perfected, this engine could operate at stoichiometric air-fuel ratios and take advantage of the advanced state of technology in exhaust-gas after-treatment for stoichiometric systems. The results of initial testing of this engine concept at AVL List GmbH have encouraged researchers to design a functional variable compression-ratio mechanism. Other novel concepts being investigated include an engine that could switch between four-stroke operation and two-stroke operation as the power demands on the engine change and a small engine that would operate at close to full load for normal operation, which is predominantly light load, and operate in "burst power" mode when high power is needed.[2]

The 4SDI technical team was also involved in the solicitation of competitive bids for the development of emission control for PNGV-sized CIDI engines. Cooperative agreements were awarded in September 1999 to Detroit Diesel Corporation and Cummins Engine Company. The 4SDI technical team supplied the committee with an estimate of the cost and affordability challenge of the proposed engine and emissions control system. The projected cost is $900 higher than the target cost of $2,100 for the combined engine and exhaust control system. The estimate of funding for collaborative research on emission control for PNGV-sized CIDI engines for 1999 was $52 million.

After-Treatment and Controls and Sensors

Meeting the new EPA Tier 2 emission standards will be a major challenge for any internal combustion engine, especially the diesel engine. Any internal combustion engine used as the primary energy converter will require substantial emissions reduction in the exhaust system. Thus, exhaust-gas after-treatment systems will be critical enabling technologies for meeting the PNGV goals. Highly effective exhaust-gas after-treatment is even more difficult for lean combustion processes, which are the most efficient processes.

PNGV has been aggressively pursuing exhaust-gas after-treatment systems in the past year. Participants in collaborative research with the national laboratories and catalyst manufacturers have included: DOE's Office of Advanced Automotive Technologies; Los Alamos National Laboratory (LANL); Oak Ridge National Laboratory (ORNL); Pacific Northwest National Laboratory (PNNL); SNL; the Low Emission Technologies Research and Development Partnership; and DaimlerChrysler, Ford, and GM. The work performed directly by PNGV and through cooperative research and development agreements (CRADAs) ranged from the fundamental development of catalysts to the testing of multicylinder engines.

SNL continued its testing of platinum-based hydrous metal oxide (Pt-based/HMO) catalysts with simulated exhaust gas and sulfur dioxide (SO_2). A patentable technology, lower light-off temperatures, and a wider temperature window for effective NO_x reductions has been transferred to the suppliers. At best, the maximum NO_x conversion efficiency was 60 percent; but a significant degradation in performance was observed when SO_2 was present in the exhaust.

---

[2] "Burst power" is a term used to describe engine operation when engine power has been temporarily increased. For example, the "burst power" could come from a significant increase in cylinder pressure charging or a jump to very high engine speed. "Burst power" would only be used temporarily and would not be a sustainable operating condition.

Suppliers sent a test catalyst with the SNL formulations to ORNL for testing on a Navistar 7.3-liter direct-injection diesel and a Volkswagen (VW) 1.9-liter turbocharged direct injection engines. There were significant differences in the conversion efficiencies of the catalyst for the two engines. For the VW engine, the conversion efficiency was less than one-half of the conversion efficiency for the Navistar engine. An evaluation showed that this difference was not related to the different levels of NO_x emitted by the two engines. Further evaluation showed that the hydrocarbon conversion in the catalyst was lower for the VW TDI engine, and there was a buildup of soot at the entry point of the catalyst. Explanations of these phenomena may be relevant to the performance differences in after-treatment systems for larger and smaller (PNGV size) engines. In any case, the tests showed that Tier 2 standards would not be attainable using Pt-based catalysts.

LANL is investigating synthesis and characterization of high-temperature supports and catalysts, including NO_x absorber technologies. One goal of these studies is to optimize the support selection, pretreatment, and catalyst selection. More than 170 materials were prepared and 150 screening tests performed. Fifty tests were made to evaluate SO_2/temperature/water stability of the materials. LANL has developed several new catalysts, filed patent applications, and is in the process of transferring the technologies to the catalyst suppliers. Although the results of these investigations have been encouraging, the catalysts are still very much in the experimental stage of development. Many questions, such as catalyst stability and overall conversion efficiencies, have not yet been addressed.

PNNL has continued its research on plasma-assisted catalysis, a process in which an electrical potential difference generates nitrogen ions that combine with NO to form molecular nitrogen and atomic oxygen. This system is used to pre-condition the exhaust gas from the engine before it enters the catalyst after-treatment system. The device was tested with simulated exhaust gas in the exhaust stream of a diesel-powered generator, and at ORNL on an engine dynamometer system. Bench test results of one catalyst showed up to 50-percent conversion of NO_x; some NO and unidentified nitrogen-containing species were detected. Results of long-term stability tests (up to eight days) and sulfur tolerance tests (50 ppm SO_2) have been very encouraging. Chemical modeling and mechanistic studies of this catalyst have been started. The results, although still in the early development stage, are promising, and the work will be continued for the next year to explore the full potential of this technology. An evaluation of combined NO_x and PM reduction using catalyst-in-plasma designs is also planned..

Another approach to eliminating PM from engine exhaust is to use a PM filter. Filter systems physically trap PM and, therefore, require a method of cleaning out the filter. Continuously regenerative traps, which require low-sulfur fuel, are being developed outside of PNGV by catalyst manufacturers. In addi-tion, in the past year, the 4SDI team has investigated the potential of using a microwave generator to initiate the burn-out of PM captured in the filter. The microwave regeneration efficiencies were evaluated under actual diesel engine operating conditions. A power of 2 kW and a time of three minutes at idle were required for regeneration. The typical engine running time between regeneration cycles was approximately one hour. This technology is considered promising, and work is planned to continue for the coming year.

Closed-loop control of the engine and emission reduction systems will be necessary to meet the new emission requirements. Once the governing fundamental interactions of the subtleties of engine operating characteristics, their resulting impact on emissions, and the performance of the emission reduction systems are understood, it will be necessary to use sensors and control strategies to minimize the exhaust emissions in all operating conditions. For example, the effects of the EGR in-cylinder distribution and the responses of the EGR system to engine transients will be important for optimizing the fuel-engine after-treatment system. Exploiting the tolerance of the engine for EGR in terms of PM and NO_x reduction will require a dynamic, closed-loop control system, for which a wide-range oxygen sensor and a dynamic PM sensor would be important enabling technologies. This is an important area for the 4SDI technical team to pursue.

At this time, the only NO_x conversion system with the requisite reduction level is a urea-selective catalytic reduction (SCR) system, which would require a urea transport infrastructure and onboard transport of the reductant. Despite the complexity of both the infrastructure and the onboard auxiliary storage system, the fact that this technology is being investigated is indicative of the monumental challenge facing PNGV.

Issues of Concern

The 4SDI technical team has made excellent progress. Nevertheless, in the short term at least, the more stringent emission standards bring into question the viability of the most efficient power plant as a primary energy converter for the PNGV vehicle. Thus, PNGV faces a dilemma: the most efficient power plant faces the most significant technical challenges in meeting the emission standards.

No doubt, lean exhaust-gas after-treatment systems will require breakthroughs to attain the conversion efficiencies required to meet the standards. In fact, new measurement systems will have to be developed to measure the extremely low PM levels mandated by the regulation. At a minimum, the after-treatment system will add complexity and probably weight to the vehicle. The operating characteristics of the engine to facilitate the after-treatment operation by maintaining the appropriate temperature and exhaust-gas composition will most likely reduce fuel economy because lean combustion exhaust-gas after-treatment systems require a reductant (usually fuel) or auxiliary power (for plasma generation). Systems models will be essential for quantifying the fuel economy penalties for various strategies of meeting the emission standards.

A second ramification of the Tier 2 emission standards is the inclusion of SUVs, which weigh up to 8,500 pounds, in the regulated fleet average emission standards. Although SUVs are not PNGV-type vehicles, the potential of improving their fuel economy with PNGV technology has not gone unnoticed. However, the very technology being considered for transfer to these vehicles (i.e., highly efficient automotive-size diesel engines) faces the daunting challenge of meeting the new emission standards. Projections by the 4SDI technical team of the after-treatment conversion efficiencies necessary for SUVs were 90 percent for NO_x and 80 percent for PM. Even though the Tier 2 standards allow some vehicles to exceed the mandated average levels, the fleet average emission level is so low that, with current and projected technologies, manufacturers may not be able to produce a sufficient number of vehicles with low enough emissions to offset emissions from SUVs. Therefore, SUV manufacturers are not likely to introduce PNGV engine technologies unless they will be able to attain the fleet average emission levels with them. Achieving an acceptable level of confidence for the emission control technology performance would require breakthroughs for lean combustion after-treatment technologies.

At this point in the PNGV program, meeting the emission targets with the CIDI engine is speculative. It seems clear that the sulfur level in the fuel will have to be reduced, but an acceptable level has not yet been quantified. Fuel composition will be critical, but compositions and properties have yet to be determined. In addition, technological breakthroughs in exhaust-gas after-treatment systems will be necessary for the most efficient energy converter to remain a viable option for a PNGV vehicle.

As PNGV continues its development activities, it will be critical to quantify and compare the fuel economy penalties associated with engine systems that are projected to meet the emission standards. An effective method to accomplish this would be to use the systems models to predict the optimal power train configuration for each engine type and after-treatment system that would meet the emission standards. The fuel economy of the different configurations could then be compared. However, before this can be done, emissions models will have to be developed and incorporated into the systems models. The development of emission models represents a critical need.

Recommendation

Recommendation. PNGV should continue its aggressive pursuit of lean combustion exhaust-gas after-treatment systems. The program should also pursue detailed systems modeling that could quantify the fuel economy penalty associated with using different technologies to meet the new Tier 2 emission standards. The modeling should also address how power train hybridization could be used to reduce emissions and what effects changing the primary energy converter would have on fuel economy trade-offs necessary to meet emissions standards.

FUEL CELLS

From the beginning of the PNGV program, fuel cells represented one (perhaps the only) energy converter technology that could simultaneously provide sufficiently low emission levels and sufficiently high energy conversion efficiencies to meet the emission and fuel economy objectives of Goal 3. However, the immature status of proton exchange membrane (PEM) fuel-cell stack technology (the technology under development for automobiles), coupled with the need for suitable subsystems (e.g., liquid-fuel processor, air-management system, etc.), virtually guaranteed that the development of the fuel cell would be on a longer time scale than most other candidate technologies considered in the PNGV program.[3]

Because of the rapid pace of achievements, as well as the potential for near-zero emissions and high energy conversion efficiency, the fuel cell has been retained as an important PNGV technology. However, because of the longer time scale, fuel cells were not considered for the "deliverable" PNGV concept cars introduced in 2000, for which the much more mature CIDI engine was selected by all three automakers as the near-term energy converter of choice. In fact, the CIDI engine in an HEV configuration was projected to yield average energy conversion efficiencies almost as high as those of hydrocarbon-fueled fuel cell vehicles, while meeting (albeit with some difficulty), the EPA emission requirements in force when the PNGV program began. However, as discussed in the previous section, the much more stringent emissions regulations have increased the challenge and cast serious doubts on the ability of the CIDI engine to meet the new emission levels in the time frame of the PNGV program. If the CIDI engine cannot meet the emission standards, and if the spark-ignited direct-injection engine cannot be modified to achieve efficiency levels close to those of the CIDI engine while meeting new emission requirements, then the fuel cell will certainly become more important for meeting PNGV's objectives. Although not considered "deliverable" concept cars, Ford has an operational version of a pressurized hydrogen P-2000 fuel cell vehicle, and GM has shown a fuel cell version of the Precept with a proposed hydride-hydrogen storage system at the North American International Automobile Show (see Chapter 4). The P-2000 fuel cell vehicle does not have the energy-saving refinements of the Ford Prodigy concept car, and the GM fuel cell Precept is not yet operational. DaimlerChrysler is also working on a fuel cell-powered version of its PNGV concept vehicle.

Three fuels remain under primary consideration for fuel cell vehicles: hydrogen, methanol, and gasoline. The primary focus in the PNGV program has been

---

[3] See previous committee reports for further discussion of fuel cell systems for automotive applications (NRC, 1997, 1998, 1999).

on gasoline. Of the three, hydrogen is the preferred fuel for the fuel cell energy converter; it allows almost instant start-up and load-following, as well as the highest system efficiency with no need for a fuel processor system. However, it is difficult to store enough hydrogen energy on board an automobile for a 350-mile range, there is virtually no infrastructure for distributing hydrogen fuel, and it is more expensive per unit of energy than the other two fuels.

Gasoline has the advantages of an extensive existing infrastructure, the highest energy storage density, and the lowest unit energy cost. However, from the standpoint of the fuel cell energy converter, it presents the greatest challenges because it requires a high-temperature fuel processor with extensive carbon monoxide (CO) "cleanup," which leads to start-up and transient delays, as well as significant efficiency losses.

Methanol falls between hydrogen and gasoline in many ways. It has an energy storage density about half that of gasoline but about five times that of hydrogen gas pressurized to 3,500 psi. It can be reformed at a much lower temperature than gasoline thus easing the start-up and transient problems some-what and providing a slightly higher system efficiency. Hydrogen is nontoxic but has by far the widest flammability limits and the lowest ignition energy of the three fuels.

At present, virtually 100 percent of both methanol and hydrogen are pro-duced from natural gas, although each could be produced from other feedstocks (although at higher costs). Methanol is already being produced at pilot plants from coal and could be produced from other materials, such as wood, grain crops, and even municipal or agricultural wastes. Hydrogen can be produced from any hydrocarbon, as well as through electrolysis of water (although it takes more energy to produce hydrogen this way than is returned as useful work by the fuel cell).

Almost all of the experimental fuel-cell vehicles currently operating use gaseous hydrogen as the fuel stored on the vehicle. A few vehicles are operating on methanol, and none is operating on fully integrated gasoline systems. The challenge of a fully integrated system is clearly related to the technical difficulties associated with using the respective fuels in efficient systems that are compatible with automobile requirements.

The committee's review this year has revealed several success stories associ-ated with R&D on fuel cell technologies. Nevertheless, some previously identi-fied problems persist, and some of the "fuzzy" problems are beginning to come into focus. The automotive companies and researchers are optimistic that the major problems will be resolved through focused R&D on evolutionary, rather than revolutionary, improvements.

Program Status and Plans

Primary activities in the past year have been focused on the following areas:

- fuel processors (including systems integration)
- stack electrodes
- air-management systems
- cost reductions

Fuel Processors

Fuel processing (converting a hydrocarbon fuel to hydrogen aboard the vehicle) is probably the most challenging technical problem facing fuel cell developers. The reformation of fuel, including gasoline, into a hydrogen-rich reformate has been demonstrated using steam, autothermal, or partial oxidation (POX) reformers. Of these, POX reformers have received the lion's share of attention (and funding) primarily because they are expected to have the shortest start-up time. Unfortunately, they also yield the smallest quantity of hydrogen per unit of fuel consumed along with copious amounts of CO, which must subsequently be removed through a combination of shift reaction to additional hydrogen and "CO cleanup" systems. These subsystems add to the start-up time (because they must also be heated), which further lowers hydrogen yield because air bled into the process gas to oxidize CO also oxidizes some of the hydrogen. These subsystems also increase costs. Even with POX processors, because of the need for shift conversion, air bleed, and preferential oxidation, the minimum start-up time from ambient temperature ranges from five to ten minutes, and the system takes a significant toll on overall energy conversion efficiency.

The major government-supported fuel processing R&D includes industry programs at Epyx Corporation/A.D. Little; McDermott Technology, Inc./Catalytica, Inc.; and Hydrogen Burner Technology, in addition to programs at Argonne National Laboratory (ANL), PNNL, and other national laboratories. R&D by Epyx/A.D. Little has resulted in a 10-kilowatt-electric (kWe) (closer to 15-kWe according to Epyx) multifueled reformer based on existing designs followed by a "new" CO cleanup section. This processor has been coupled with a 10-kW Plug Power, Inc., PEM stack. Once steady-state performance was satisfactory, the fuel processor/stack combination was tested with gasoline and several other alternative fuels at various steady-state operating conditions.

The tests demonstrated the operation of a stack for relatively long periods of time (hundreds of hours) at various steady-state power levels fueled by reformate obtained through an efficient conversion of a variety of hydrocarbon fuels. Also, steady-state emission levels were far below EPA Tier 2 standards. However, the start-up time for the fuel processor from ambient temperature was about seven minutes, the test processor was not an "integrated" system, and the stack and fuel processor were operated separately. No system transient capability was shown, and the relatively high projected federal test procedure cycle emission levels (based on start-up and limited transient processor data) indicated that a new tail-gas combustor, which utilizes gas discharged from the stack, would be necessary.

As part of the same Epyx/A.D. Little project, a 50-kWe fuel processing system is under development that will be integrated into a 50-kWe Plug Power, Inc., fuel cell stack. The intent is to create an integrated system with interactions and feedbacks between fuel processor and stack that can be studied in detail. The 50-kWe integrated system is expected to be operational by the summer of 2000.

Two other PNGV-sponsored industrial projects are under way, both to develop 50-kWe fuel-flexible processors. The McDermott Technology project was begun in late 1999, and the Hydrogen Burner Technology project is in the process of "performance mapping" to characterize the system.

As noted in the committee's fourth report, significant private sector investments are also being made worldwide in fuel cell programs for vehicles, and substantial analytical and experimental capabilities are being developed by private industry for long-range development of fuel cell systems (NRC, 1998). Additional activities have been initiated recently. For example, International Fuel Cells (IFC) has reorganized its existing facility in South Windsor, Connecticut, and added new leased space to house a PEM system development project. New research facilities are being dedicated at IFC to PEM stack development, as well as fuel processing for hydrogen, methanol, and petroleum-based hydrocarbon fuels.

Another impressive industrial fuel cell development project is under way at the GM Global Alternative Propulsion Center (GAPC) near Rochester, New York. In addition, there is an operation of similar scope in Mainz-Kastel, Germany, and a smaller operation at the GM Technical Center in Warren, Michigan. Like IFC, GAPC has virtually all new R&D facilities devoted to the development of PEM fuel cell components and systems. GAPC is also working with hydrogen, methanol, and hydrocarbon fuels.

In addition to the projects sponsored by PNGV, IFC has delivered three Series 200 ambient-pressure 50-kW units to undisclosed automotive companies. Presumably these units are configured to operate with hydrogen fuel. (IFC is also working with at least two other automotive companies.) IFC has delivered a 5-kW system to BMW as an auxiliary power unit that also uses hydrogen (liquid) fuel. Apparently, IFC has also successfully operated Series 200 units with gasoline over a range of conditions, including transients (although not as a fully integrated system), and is currently developing the Series 300, which is expected to have better overall efficiency, a faster start-up time (two to four minutes), and better specific power. IFC is also working on Series X for the 2005 time frame, which is expected to meet or approximate PNGV 2004 performance and cost targets.

GAPC has successfully operated (and continues to operate) a fully integrated 30-kW methanol system. It has also developed and is currently operating all of the major components of a complete gasoline-fueled PEM system. Although this system is not yet operating as a fully integrated unit, full integration is expected very soon.

GAPC has taken a comprehensive approach that involves joint activities with dozens of potential suppliers. "Four win" criteria have been developed for product success: (1) the customers (performance, cost, safety, etc.); (2) GM and automotive suppliers; (3) energy companies; and (4) society at large (low emissions, fuel conservation, etc). GAPC indicates that, although some serious problems are still unresolved, the technology is expected to be ready by 2004. Large-scale production will probably not be feasible for an additional four to six years.

All of the major processor, stack, and systems development projects are currently working toward approximately 50 kW as a nominal design power level. This power level, which is insufficient for a nonhybrid vehicle configuration, is based on the assumption that liquid-fueled systems will require hybridization to shorten start-up times, transient response times, and even overcome cost barriers, a correct assumption based on the current state of the technology. However, hybridization also represents increased vehicle power-train complexity and weight. Therefore, as the cost per kilowatt for fuel cell systems declines and system response improves, the hybridization is likely to be eliminated and the design power level significantly increased (perhaps doubled or more) by the developers.

Two of the national laboratory programs have shown promising results to date. ANL has developed a new catalyst that can operate at higher temperatures (~700°C) than conventional catalysts and does not have to be protected from exposure to air during or after shutdown. These characteristics could be beneficial in reducing processor size, complexity, and start-up time in an autothermal reformer.

PNNL has successfully used microchannel technology to develop ultra-compact reactors and heat exchangers. The results of their efforts have revived interest in steam reforming, which is more efficient than alternative reformer concepts but was believed to be incompatible with automotive applications because of limited heat transfer rates. Results to date appear to confirm the potential for dramatic reductions in fuel processor size and to show fuel conversions (with iso-octane) approaching 90 percent. However, fuel conversions will have to be much higher (> 99 percent), or a technique for separating out the unreacted fuel will have to be used, to avoid damaging the stack.

Stack Electrodes

Efforts to improve stack electrodes are generally focused on the following objectives:

- increasing CO tolerance
- reducing precious-metal loading
- improving manufacturability

Increased CO tolerance equates to lower demands on the fuel processing cleanup sections with corresponding increases in system efficiency (reduced fuel consumption) and decreases in processor weight, volume, cost, and start-up time. To date, the most successful efforts have come either through heavier precious-metal loadings (with corresponding higher electrode costs) or through a higher cell operating temperature (with corresponding durability and life reductions using currently available membrane materials).

Efforts to increase CO tolerance, primarily at LANL, have been focused on improving alloy electrocatalysts, air bleeding, anode design, and increasing stack temperatures. Although CO tolerance has been increased with these technologies, they require either precious-metal loadings (higher costs) or reduced efficiency (air bleed). Operation at an elevated temperature (120°C to 150°C) is likely to mitigate these problems but will require pressurization to maintain water content; in addition, no known membrane materials have adequate lifetimes at elevated temperatures.

Reducing precious-metal loading was a major goal when loadings were as much as an order of magnitude greater than they are with current technology; at that point corresponding high projected costs (as well as availability) were potential showstoppers. As loadings were dramatically reduced, membrane and projected manufacturing costs became priority issues. Now, as projected membrane costs have plummeted and the feasibility of low-cost molded plates and high-volume production of membrane electrode assemblies have been demonstrated, precious-metal loading has again become a limiting cost factor (although much less than it was).

Efforts to reduce precious-metal loadings have been directed mostly toward alternative catalysts and improving the use (surface structure) of precious-metal catalysts. R&D, primarily at LANL, has resulted in continuing improvements in the past year but no major breakthroughs. The noble-metal requirement for the fuel cell stack, based on current technology, is more than an order of magnitude greater than for current automobile catalytic converters. If noble metal is also used in the fuel processor, substantially more would be required. The lack of sufficient noble metal could be a problem for large-scale production.

Air-Management Systems

Air management has evolved into a high-priority issue because the first phases of developmental contracts by five organizations have not produced a single system that meets all of the important targets simultaneously. Smaller high-speed centrifugal machines do not provide the necessary efficiency or pressures at reduced airflows (reduced power output). Variable-displacement machines and "scroll" designs have performed well over a range of rates but are large and bulky. Twin-screw devices provide satisfactory pressure and airflow but are noisy and inefficient. The remainder of work in this area is focused on developing a novel air bearing for the rotating parts.

An independent panel sponsored by ANL to review compressor/expander technologies is working with DOE to study the air-management problem and current R&D. Meanwhile, various developers are continuing to improve the current technologies. In addition, work is continuing at IFC, national laboratories, and, probably, all major private sector developers on ambient-pressure PEM fuel cell systems that would eliminate the need for compressor/expander subsystems.

One of the issues coming into focus through more detailed system studies is high overall-mission thermal efficiency. A LANL study showed that to achieve the 2004 target of 48 percent overall system efficiency at 25 percent of peak power (where most operation is expected), will require a cell voltage of 0.9 V (Milliken, 1999). Therefore, the cell voltage at rated power (50 kW) would be about 0.77 V, which is much higher than was previously assumed (about 0.55 to 0.60 V). Unfortunately, the power density of the stack is much lower (perhaps 30 percent lower) at the higher voltage, thus implying the need for a larger stack. Therefore, stack efficiency will have to be increased, most likely through improvements in cathode performance. Experience also highlights the need for more and better systems analysis.

Cost Reduction

A cost study by A.D. Little based on apparently realistic assumptions and state-of-the-art technology puts the projected high-volume "factory cost" of the complete automotive (hydrocarbon fuel) fuel cell system at $294/kW, more than double the 2000 PNGV goal of $130/kW (Milliken, 1999). The stack was about half of the projected cost ($140/kW); precious metals in the stack accounted for almost $60/kW. Another concern is that projected demands for platinum (and other precious metals) could exceed the world supply if fuel cell vehicles eventually replace internal combustion engine vehicles. A similar concern was raised in the 1970s when catalytic converters were first introduced. This worry turned out to be a nonissue, however, as improvements in technology dramatically reduced the demand for precious metals, and supplies were increased. It is too early to tell if similar developments would occur with fuel cells.

Areas of Concern

Even though efforts are continuing or being initiated in all essential areas, the following significant issues must be addressed:

- Costs. Even though projected costs have been reduced dramatically in the past few years, they are still approximately double the PNGV target for 2000 and about six times too high for a cost-competitive power plant.

- Lack of an integrated system. Hydrocarbon fuel processors have supplied hydrogen-rich fuel gas to PEM stacks, but the only complete *integrated* systems have been fueled by hydrogen or methanol, neither of which may be acceptable in the marketplace because of consumer resistance or infrastructure problems. This is a critical issue because liquid hydrocarbon fuels (e.g., gasoline) are believed to be extremely important for consumer acceptance in the marketplace.
- Fuel processing systems. The best efforts to date have not resulted in a fuel processing system that can meet the required start-up times. The demonstrated systems do not have attributes (e.g., size, shape, weight, cost) that would be compatible with production vehicles. In addition, no fuel processing system has been successfully integrated into a complete fuel cell "engine."
- Systems studies and trade-offs. Because consistent targets are necessary for orderly technological development, the best analysis tools and available information should be used to determine the relative importance of apparently conflicting targets. For example, how important is the cell voltage required to meet efficiency targets at 25 percent of rated power compared to the weight, volume, and cost targets? Similar questions could be asked about the fuel processor. Assuming that all targets cannot be met, what are the relative penalties for meeting or not meeting each one in relation to the others?
- Air management. None of the technologies under development appears to be capable of meeting all targets simultaneously, which clearly suggests that an ambient-pressure system should be pursued. However, with ambient-pressure systems, meeting stack efficiency targets is likely to result in even lower power densities. Thus, more emphasis should be placed on efficient pressurization systems.

In summary, fuel cells will remain very important to the PNGV program—especially as Tier 2 emission regulations are implemented. Significant technical progress continues to be made, but some important technical issues and many cost issues persist. The emergence of large-scale industry efforts by both system developers and potential suppliers in the United States will undoubtedly hasten the reality of fuel cell vehicles for consumers and should be considered a major accomplishment of the PNGV program. PNGV's experience with fuel cell development suggests that DOE should return to supporting high-risk, high-payoff R&D and focus less on supporting hardware development.

Recommendations

Recommendation. In the area of fuel cell development, PNGV, and especially the U.S. Department of Energy, should emphasize high-risk, high-payoff research in critical areas, such as fuel processing, carbon monoxide-tolerant electrodes, and air-management systems.

Recommendation. PNGV should conduct trade-off analyses to establish relative priorities for fuel cell technical targets and cost targets.

Recommendation. PNGV should consider conducting a comprehensive assessment of the consequences of fuel choices for fuel cells and their impact on PNGV's direction and ultimate goals.

ELECTROCHEMICAL ENERGY STORAGE

In the simplest HEV drive train, the battery stores vehicle kinetic energy captured in regenerative braking and provides stored energy to start the vehicle's combustion engine and assist the engine during acceleration. Overall vehicle energy efficiency compared to a conventional vehicle is enhanced if braking energy can be recovered and if the engine can be operated at a nearly constant power level. This is the basis for the HEV, which stores energy for use in various parts of the drive cycle. Energy storage can also compensate for a slow-response engine, such as a fuel cell with a fuel processor. Some energy storage systems may allow an HEV to operate as a zero-emission vehicle that can run on stored energy alone for a certain distance. Also, the emissions and fuel economy with an internal combustion engine should be improved if the engine load varies over only a narrow range. Batteries designed for high specific power are the most likely means of energy storage. PNGV eliminated flywheels and electrochemical capacitors for energy storage in HEVs during the technology selection process in 1997.

The HEV power train chosen by most automotive manufacturers engaged in HEV development requires a battery with relatively small energy storage but very high power capability. At the outset of the PNGV program, the goals for battery specific power and power density were far above what was commercially available. Table 2-1 summarizes the battery performance and cost targets for the low energy storage power-assist mode and the dual mode. Power-assist and dual modes refer to different degrees of hybridization. In the power-assist mode, the power goes directly from the primary power plant to the wheels, with some power peaks being supplied by the energy-storage subsystem. In the dual mode, the primary power plant is run substantially at a constant power level, with all fluctuations being supplied by the energy-storage system (see the committee's fourth and fifth reports, for more extensive discussion [NRC, 1998, 1999]). The battery in a dual-mode HEV can provide the energy for propelling the vehicle while the slow-response energy converter (e.g., a fuel cell with a fuel processor) is warming up.

TABLE 2-1 Design Targets and Current Performance for Short-Term Energy Storage

	PNGV Targets		Current Performance	
	Power Assist	Dual Mode	Li-ion[a]	NiMH[b]
Pulse discharge (constant for 18 s) (kW)	25	40[a]		
Maximum regenerative pulse power				
(at start of a 10 s trapezoidal pulse) (kW)	30	40		
Available energy (kWh)	0.3	1.5[a]		
Energy efficiency (%)	> 90	> 95	90	88
Calendar life (years)	10	10	3	5
Maximum mass (kg)	40	65		
Maximum volume (L)	32	40		
Production costs @ 100 k/y ($)	300	500		
Minimum operating temperature (°C)	−40	−40	−25	ok[d]
Maximum operating temperature (°C)	52	52	50	ok[d]
Available specific energy (Wh/kg)	7.5	23.1	23	15
Specific power (18 s) (W/kg)	625	615	625	400
Cost/available energy ($/kWh)	1,000	333	1,041	1,064
Available energy density (Wh/L)	9.4	37.5	29	23
Cumulative contracts[c] ($ millions)			55	8.1

[a]Revised downward during the year.
[b]Cost, specific energy, and energy density refer to dual mode.
[c]Includes 50% cost sharing on industrial contracts.
[d]Operates at the required temperatures.

Source: PNGV, 1999a; Haskins, 1999.

PNGV is developing both lithium and nickel metal hydride (NiMH) batteries for HEVs. NiMH cells and modules for HEVs have reached technological maturity but fall somewhat short of meeting the power-assist targets. Hopes have been pinned on lithium or lithium-ion (Li-ion) batteries for high energy and energy efficiency, but NiMH batteries are still a less risky backup system for nearer term deployment in concept vehicles. The approximate status of these battery systems in terms of the PNGV goals is shown in Table 2-1.

A key requirement and development target is the capability of the hybrid battery to provide an available energy of 300 Wh while weighing no more than 40 kg. Available energy is energy that can be delivered while simultaneously meeting the pulse-power targets for discharge and regeneration (acceleration of a 1,000 kg vehicle to 100 km/h requires 0.1 kWh for the kinetic energy alone). Meeting the cost and life goals remains a major challenge.

Program Status and Plans

Accomplishments

The life projection has been improved to three years for the Li-ion system, but it is still based on an accelerated testing schedule. As understanding of the HEV system has improved, the goals for the battery have been modified. For example, the PNGV targets for available specific energy have been reduced compared to last year's targets (NRC, 1999). Some goals on the cell level are now the same for the dual mode and the power-assist mode, which means that developers can concentrate on one cell design. Because the power/energy ratio is different for the two modes, a larger range of state-of-charge can perhaps be used in the dual mode; that is, the PNGV target "energy utilization factor," the ratio of available energy to rated energy, can be 0.378 for the dual mode as compared to 0.141 for the power-assist mode. However, the required specific power is still the same for the two applications. Goals have also been modified to allow for power fade during the life of the battery.

The SAFT Li-ion program and the PolyStor Li-ion program, which have been ongoing for one or more years, include the development of 50-V modules. A VARTA Li-ion program has been terminated, but a number of new contracts involving new developers for lithium battery development for HEVs have been initiated. Hydro-Québec, which is working on a lithium polymer system, and Delphi, which has a Li-ion, gelled-electrolyte system, have been added. VARTA, the principal developer of the NiMH system has delivered and tested 50-V modules. (The battery in Ford's concept vehicle is based on this technology.) GM-Ovonic has been added to develop an NiMH system. Another small project by Electro Energy to explore the potential of a bipolar NiMH system has also been initiated.

The Advanced Technology Development program at the national laboratories is looking into the root causes of cell failure of Li-ion batteries. Substantial resources have been devoted to producing test cells of about 1 Ah with non-proprietary chemistry. These cells are tested for different regimes of cycling and temperature and subsequently examined by various diagnostic techniques (including calorimetry, microscopy, and impedance as well as techniques for documenting chemical and structural changes in electrodes) to identify the inherent causes of cell failure and capacity fading. The program also addresses approaches, such as improved materials and electrochemistry and low-cost packaging, for reducing cell costs. Developers are working on safety and the control and balance of series-connected strings of cells.

A detailed cost analysis was performed on the more developed systems to provide a firmer basis for cost projections and to indicate how costs might be reduced. The cost of both the NiMH and Li-ion systems exceeds the PNGV cost goals by about a factor of three.

Assessment of the Program

The modules and cells developed as a result of PNGV joint activities with the battery developers are not necessarily being used in vehicles. Nevertheless, the program has advanced HEV battery technology substantially. The automotive companies have made their own arrangements with battery suppliers for systems in their concept vehicles, which enables them to use systems analysis to coordinate the battery technology requirements and performance with other parts of the vehicle in a proprietary setting. As a consequence, the PNGV program should focus on generic issues, such as producing systems with good performance and dealing with issues of life, cost, and safety, as well as control of series strings and thermal management.

Safety is still a concern with Li-ion systems. Overheating leads to substantial electrolyte venting, and flames are difficult to extinguish. Projected costs are too high by a factor of three, and calendar life is a serious issue.

The NiMH system has longer life, largely avoids safety problems, and should be more benign in control and balance of series strings. Projected costs are still too high by a factor of three, and the system does not provide quite enough available energy or power.

International Developments

Japanese battery companies are the world leaders in the development and commercialization of advanced high-power batteries for HEVs. Their leadership, based on the technology and manufacturing basis established over the past decade in NiMH and Li-ion batteries for consumer applications, is now driven by competition for the emerging HEV battery market represented by the Toyota Prius and Honda Insight HEVs.

Nickel Metal Hydride Batteries

Panasonic Electric Vehicle Energy (PEVE) established the world's first commercial production of an HEV battery. PEVE's HEV-6.5 technology has been used in the Prius as a 1.9-kWh, 22-kW battery weighing 44 kg; performance falls about 20 percent short of the key specific power target of PNGV. Recently, PEVE completed the development and established a production line for advanced-design NiMH HEV battery modules. Prius HEVs will soon be equipped with 40-module, 40-kg batteries that will deliver pulse power of more than 30 kW, thus exceeding the PNGV performance target. These batteries are likely to meet PNGV cycle-life targets and will have significantly improved life at elevated temperatures. Also, in mass production, the cost will be below the approximately $1,000/kW for the first-generation Prius battery.

Lithium-Ion Batteries

Other leading Japanese battery companies are concentrating on the development of Li-ion batteries for HEV applications. Sony, Shin-Kobe (part of Hitachi), and JSB have pilot plants for Li-ion cell fabrication and are offering preprototype cells and/or modules for evaluation. All of these technologies meet or exceed PNGV's specific power targets and are claimed to have very long cycle life over narrow ranges of depth of discharge; calendar life is not fully established but is likely to be much longer than three years if the batteries are operated in the HEV mode and at temperatures below about 45°C. The positive electrode in all of these technologies is based on lithium manganese oxide spinel, the lowest cost material among currently established positive materials. It is not clear at present whether the cost of mass-produced Li-ion HEV batteries would be lower than the cost of NiMH batteries. (The unit costs projected for both Li-ion and NiMH batteries are well above the PNGV targets.)

Recommendation

Recommendation. PNGV should continue to work on cell chemistry of lithium-battery systems to extend life and improve safety, while continuing to lower costs. PNGV should continue to refine performance and cost targets as overall vehicle systems analysis determines the optimal degree of vehicle hybridization. PNGV should also support efforts to apply the materials improvements achieved in the program to improve lithium battery technology and to validate improved performance, life, and safety.

POWER ELECTRONICS AND ELECTRICAL SYSTEMS

The technology selection (downselect) process in 1997 resulted in all three USCAR partners adopting an HEV configuration for their PNGV vehicles. The cost, volume, complexity, and weight of the electrical and electronic systems in an HEV are substantially greater than in a conventional, nonhybrid car powered by an internal combustion engine. Consequently, the chances of an HEV design achieving the PNGV cost and fuel economy goals depends to a great extent on the power electronics and electrical systems meeting their targets. Table 2-2 shows the present values and the 2004 PNGV target values (Malcolm et al., 1999). The targets are extremely aggressive, and meeting them will require breakthrough developments in both component and manufacturing technologies.

Program Status and Progress

The Electrical and Electronics Systems Technical Team (EE Tech Team) has implemented an aggressive and diverse R&D program to meet the electrical and

TABLE 2-2 Current Specifications and Target Specifications for Power
Electronics

Component	Specific Power (kW/kg)	Efficiency	Cost ($/kW)
Power electronics			
Today	4 kW/kg	95%	$10/kW
2004 target	5 kW/kg	97–98%	$7/kW
Motor/generator			
Today	1.5 kW/kg	92%	$6/kW
2004 target	1.6 kW/kg	96%	$4/kW

Source: Malcolm et al., 1999.

electronic systems targets. The committee compliments the EE Tech Team on the comprehensiveness and relevance of its program, as well as its organization and management. The program has three elements: power electronics development, electric motor development, and system development.

The power electronics component of the program includes the development of an automotive integrated power module (AIPM), the development of new thermal management techniques and materials, and the investigation of new materials and manufacturing processes for passive energy-storage components, particularly capacitors. Three contracts have been awarded for the development of an AIPM, and each of the contractors—Semikron, Silicon Power Corporation (SPCO), and SatCon—has projected that the PNGV cost targets can be met. Semikron expects to employ existing power electronics technology in a high-volume, low-cost process. SPCO is focusing on unique packaging and silicon device technology. SatCon intends to meet the AIPM objectives by employing flip-chip packaging technology and creative thermal management designs.

The national laboratories have a large number of ongoing research programs on dielectric materials and capacitors. The EE Tech Team has reviewed these programs and is in the process of developing a road map to meet PNGV's passive component needs by leveraging these programs.

The programs on electric motors are focused primarily on investigating materials and manufacturing processes that could reduce costs. Other activities include the assessment of accessory drives[4] and studies of different machine types that could be made practical by low-cost power electronics, such as switched reluctance and high-speed machine designs. A major element of this part of the program is the development and demonstration of an automotive electric motor drive (AEMD). The goal is to use the AIPM as part of a complete motor drive

---

[4] The accessory system comprises electrically or mechanically driven devices that are not part of the drive train (e.g., air conditioner, power windows, ventilating fans, power steering, etc.).

system. Two teams have been selected for this work, one led by Delco Remy International and the other by Delphi Automotive Systems. The Delphi team has proposed using the AIPM in combination with an alternating-current (AC) induction motor; the Delco Remy team will use a direct-current (DC) brushless motor.

Projects on system development are focusing on integrated packaging of the power electronics and machine, as well as the modeling of system cost.

Assessment of the Program

The EE Tech Team has done an excellent job of identifying the key barriers to meeting the PNGV cost and performance targets for the electrical and electronic systems and of leveraging ongoing work by industry, universities, and the national laboratories.

As the committee has stated previously, the functional specifications of the power electronics and electrical systems have been met (NRC, 1999), and the remaining challenges include meeting the physical (packaging), efficiency, and, particularly, cost targets.

In its fifth report, the committee expressed its concern about the PNGV's prospects for meeting its cost targets for power electronics and motors. The extensive discussion of this issue will not be repeated here (NRC, 1999). The contractors for the development of AIPM and AEMD all claim they can meet or exceed the cost targets, but the EE Tech Team and the committee have been shown little supporting data for these claims. The committee was shown drawings of proposed AIPM designs but no detailed cost analyses. The committee reviewed production cost goals for SatCon's proposed design, which showed that the current cost of $20 to $30/kW is expected to be reduced to $5 to $9/kW in the final production prototype. This implies a substantial reduction in component cost, as well as manufacturing process cost. However, this projection is predicated on developments by component suppliers, whose level of confidence regarding the success of these developments was not clear. For example, the EE Tech Team expects that the cost of the high-voltage bus capacitors will be reduced from $0.20/microfarad ($\mu$F) to $0.02/\mu$F but did not explain how this would be achieved (Malcolm et al., 1999).

In fairness to the EE Tech Team, many cost projections must be taken on faith because they depend on developments that are under way but not yet demonstrated. However, given the importance of the electrical and electronic systems to the viability of the entire PNGV vehicle, the EE Tech Team must always be aware of progress toward cost goals and must communicate this information to the systems analysis team.

In its fifth report, the committee recommended that the EE Tech Team perform an analysis of electrical accessory loads to verify system needs. The vehicle engineering team has taken the responsibility for reducing energy consumption at the vehicle system level, and the EE Tech Team intends to initiate a program to develop efficient motors for accessories. However, to the committee's knowledge a system level study of electrical requirements has not yet been done.

Recommendations

Recommendation. The Electrical and Electronics Systems Technical Team should closely monitor the progress toward meeting the cost goals of the automotive integrated power module and automotive electric motor drive and update and communicate realistic expectations for costs in 2004 to the systems-analysis team.

Recommendation. The Electrical and Electronics Systems Technical Team, in collaboration with the vehicle engineering team, should undertake a comprehensive study to identify the electrical load requirements of the accessory system. Although the details of the accessory system will differ among the three USCAR partners, the impact of the accessory load is important enough that it should be considered explicitly by the systems analysis team.

STRUCTURAL MATERIALS AND SAFETY

The reduction of vehicle mass through improved design, lightweight materials, and new manufacturing techniques is one of the key strategic approaches to meeting the PNGV Goal 3 fuel economy target. To achieve the 80 mpg target, systems analyses showed that a 40-percent reduction in vehicle weight would be necessary, together with: 40 to 45 percent power train energy conversion efficiency, 70 percent efficiency[5] for regenerative braking, improved driveline efficiency, and reduced aerodynamic drag. At the same time, the baseline vehicle performance, size, utility, and affordable cost of ownership must be maintained.

The substantial vehicle weight reduction targets for various subsystems (Table 2-3) would result in an overall reduction in curb weight of 1,200 lbs (40 percent), and a 2,000 lb vehicle.

Materials Selection, Design, and Manufacturing

In its search for lightweight materials that would enable the targeted large weight reductions (50 percent for the body-in-white [BIW],[6] for example), PNGV has focused on materials with densities substantially lower than the density of

---

[5] The regenerative braking efficiency is the proportion of total braking energy that is returned to the vehicle as useful propulsion energy.

[6] Body-in-white (BIW) includes all the structural components of the body, the roof panel, and the subframes, but not the closure panels.

TABLE 2-3 Weight-Reduction Targets for the Goal 3 Vehicle

Subsystem	Current Vehicle (lbs)	PNGV Vehicle Target (lbs)	Mass Reduction (%)
Body	1,134	566	50
Body-in-White	590		
Chassis	1,101	550	50
Power train	868	781	10
Fuel/other	137	63	55
Curb weight	3,240	1,960	40

Source: Adapted from Stuef, 1997.

steels used in the baseline vehicle (NRC, 1998, 1999). Components and BIW structures fabricated primarily from aluminum, glass fiber-reinforced polymer composites (GFRP), and carbon fiber-reinforced polymer composites (CFRP), including hybrid structures (NRC, 1999), are being investigated. Lower density materials entail an incremental increase in cost compared to the baseline steel BIW and closure panels, probably on the order of $1,400 (Schultz, 1999), depending on several variables. The cost penalty for CFRP BIW would be substantially higher.

Another approach to weight reduction has been taken by the steel industry. A consortium of 35 producers of sheet steel was formed in September 1995, under the auspices of the American Iron and Steel Institute, to design and build a prototype of an ultralight steel auto body (ULSAB) (Jeannes and van Schaik, 2000; USLAB, 1999). This consortium is not part of the PNGV program. Porsche Engineering Services carried out the engineering analyses and manufacturing management of the project. The development of the ULSAB BIW, completed in 1999, cost $22 million over four years.

In carrying out this project, the steel industry investigated a class of steel grades with varying strength but with essentially the same density. They decided to explore the weight reductions that could be achieved through: (1) greater use of higher strength steels than in the baseline vehicle and steel/plastic sandwich structures; (2) finite element method (FEM) modeling for performance, particularly for torsional and bending stiffness and crash performance; and (3) innovative manufacturing processes, such as laser-welded tailored blanks and hydroformed tube structures and roof panels.

All of these processes were integrated into a unified approach to reduce the average thickness of steel sheet and thereby save weight. At the end of the study, the ULSAB BIW weighed only 447 lbs, a 24-percent reduction over the PNGV baseline vehicle described in Table 2-3, and as much as 36 percent lower in weight than the heaviest PNGV vehicle benchmarked in the ULSAB study. Because the ULSAB study is very well documented, the PNGV partners should be able to verify the results (USLAB, 1999). The study also concluded that the ULSAB BIW would cost $154 less than the baseline PNGV BIW.

The American Iron and Steel Institute has embarked on a follow-on study to ULSAB, named ULSAB-AVC (advanced vehicle concepts), which will result in complete design concepts for an ultralight steel-intensive car that meets 2004 vehicle and crash requirements, including closures, suspensions, engine cradle, and all structural and safety-relevant components, for which steel offers mass and/or cost-efficient solutions. The ULSAB-AVC study, which will be completed in early 2001, will be based entirely on computer modeling. If the modeling results are promising, the American Iron and Steel Institute may build a prototype in the next phase of development (USLAB, 1999), which would be a PNGV-type vehicle (i.e. a five-passenger, four-door sedan, with an overall length of 187 in [4,750 mm] and a total weight of 2,275 lbs [1,032 kg] when powered by a gasoline-fueled internal combustion engine).

The ULSAB-AVC study has adopted an interesting approach to achieving the low total weight of 2,275 lbs. The BIW weight is assumed to be 447 lbs (203 kg), the level achieved in the ULSAB prototype. Actually, crash standards in 2004 will require that 55 lbs (25 kg) in new or strengthened structures be added to the BIW. Therefore, to maintain a 447 lb BIW, design efficiencies would have to be found in BIW components that are less involved in crash performance. As pointed out earlier, a 447 lb BIW represents a 24-percent improvement over the PNGV baseline vehicle, whereas a 50-percent reduction is required (Table 2-3). The shortfall of 152 lbs (69 kg) would have to be offset by additional reductions in other subsystems, such as closures, chassis, and the power train. The study will identify potential reductions and investigate their performance through modeling. Thus far, the ULSAB-AVC concept does not include a hybrid diesel/electric power train, which will make it difficult to make direct comparisons between the efficient steel concept and the aluminum-intensive vehicle approach.

Vehicle Crashworthiness

The crashworthiness of lightweight vehicles, such as the PNGV concept cars, depends significantly on the design innovations and lightweight materials used in the crash-energy-absorbing structures of the vehicle body. Increasingly stringent crash standards usually result in heavier body weights. For example, the ULSAB BIW was designed to meet future crash requirements, such as 35 mph frontal and rear impacts (compared to the 30 mph impacts for current federal motor vehicle safety standards). These standards were not in place in 1994 when the PNGV benchmark vehicles (midsized family sedans) were selected. The more stringent standards will require that 55 lbs be added to the ULSAB 447-lb BIW.

Current crash design practices based on barrier crash requirements (NRC, 1999) have used low-stiffness front ends to ensure that decelerations are in the range of l5g to 25g. If a lightweight, full-sized vehicle, such as the PNGV Goal 3 vehicle, followed current design practices, it would be at a disadvantage in crashes with heavier cars because the soft front end would cause most of the crash energy to be absorbed by the lighter car while significant deformation of the heavier car had barely begun. To ensure compatibility in car-to-car collisions, it has been argued that light and heavy cars should crush at the same load level, which implies that the deceleration of the lighter car would have to be faster than for the heavier car by the ratio of the mass of the heavier car to the lighter car (Frei et al., 1997). Of course, the resisting force of the lighter car structure could be designed so that the deceleration does not exceed a safe level of 40g to 50g. With this approach, the crush distance of each vehicle would be proportional to the mass of the vehicle. The crush zone of the lighter car would have to be long enough to absorb at least its own kinetic energy. These concepts have been confirmed in car-to-car crash tests (Frei et al., 1997; Niederer, 1993).

Proprietary crash modeling is being used extensively by the USCAR partners to develop crashworthy 2000 concept vehicles. The PNGV program, however, is exploring advanced crash modeling methods at ORNL by running modern crash models on high-speed, parallel computers to develop crashworthy structures of steel, aluminum, and composites, as well as hybrid-material structures (Simunovic and Carpenter, 1999). The ORNL models are state of the art and include materials-related effects, such as strain rate sensitivity, and materials processing effects, such as springback.

ONRL has built complex FEM models of several vehicles: ULSAB BIW, a Ford Taurus, an AUDI A8, and a Ford Explorer. The models are sufficiently complete so that the exterior can be peeled away to reveal interior components, and sectional views can also be taken. The primary use of these models will be to simulate car-to-car collisions and barrier crashes.

These studies should be continued and used in developing design practices to mitigate the disadvantage of lightweight cars in collisions with heavier vehicles. Recent studies are addressing such issues, for example, Evans (2000) has empirically analyzed the on-highway effects of changing car mass without varying car size from National Highway Traffic and Safety data. In addition, the safety implications of the sensitivity of lighter vehicles to wind gusts should be considered in the design of PNGV vehicles.

Materials Road Map

The PNGV materials team has developed a materials road map identifying lightweight material alternatives for the major subsystems of the vehicle (PNGV, 1999a). Alternative materials are prioritized based on total weight saving potential and feasibility to meet the PNGV time frame requirements. Technical challenges are also identified and prioritized by the likelihood that promising solutions will be found to overcome the challenges. The road map, which was summarized in the committee's fifth report, is essentially the same this year (NRC, 1999).

More than 40 materials projects have been initiated to tackle the technical challenges identified in the materials road map. The committee believes that these projects have the potential to remove the barriers to the use of alternative materials in current vehicles. The requirement that the overall cost of vehicle ownership not be increased presents a major hurdle to meeting the fuel economy goal. The road map reflects this requirement by defining reduction in the cost of feedstock as a common challenge for all alternative materials. The committee is pleased to note that PNGV personnel are working closely with materials suppliers to develop less costly manufacturing processes and new design practices that use materials more efficiently.

Program Status

Body System

The existing baseline vehicle body system weighs approximately 1,100 lbs (500 kg) and is targeted for a 50-percent weight reduction. Because the BIW (590 lbs [268 kg]) and the closure panels (220 lbs [100 kg]) account for a large fraction of the weight, the PNGV team has concentrated on these components and has identified several alternatives: the ULSAB steel-efficient concept, aluminum sheet and extrusions, fiber-reinforced polymer composites with GFRP or CFRP reinforcements, and hybrid material concepts (e.g., an aluminum space frame that supports polymer composite panels).

Steel. The ULSAB steel-efficient designs for a high-strength steel BIW (discussed above) have demonstrated a 24-percent weight saving over the baseline vehicle in Table 2-3. ULSAB is carrying out a modeling study to reach a steel-intensive vehicle weighing 2,275 lbs (1,033 kg), which may be further reduced by hybridizing with aluminum or other low-density materials.

Aluminum. The low-cost aluminum sheet project to develop 5000-Series continuous cast sheet for body structures has been completed successfully. Belt-cast thin sheet, using several process/alloy variations, has been used to form several large and challenging prototype parts for PNGV.

Progress is also being made in aluminum manufacturing and assembly. An apparatus has been built to control the binder force during the forming stroke, which will increase the percentage of good parts made. To improve the formability of aluminum, warm forming trials are being conducted. Another project to evaluate high-pressure water pulse joining has been initiated. A project to study laser welding of aluminum stamping blanks with different thicknesses and/or strengths (tailor-welded blanks) has been completed and a follow-on project is planned. An aluminum alloy scrap-sorting project has also been initiated.

Polymer Composites. A lightweight hybrid body employing bonded CFRP and aluminum structures successfully achieved a 68.5-percent weight reduction over the baseline vehicle (NRC, 1999). In a successful 30-mph frontal crash test, this lightweight body exhibited less deceleration than the baseline vehicle.

Projects are in place with suppliers and ORNL to develop low-cost carbon fibers through lower cost precursors and by microwave processing. Costs could be reduced by as much as 20 percent.

DaimlerChrysler has been exploring injection molding of large integrated moldings for body panels from thermoplastic resins, both nonreinforced resins (e.g., polyethylene-terephthalate) and glass-reinforced thermoplastic resins. DaimlerChrysler has reported building a Composites Concept Vehicle and a concept Jeep Commander from such materials and believes it can achieve 50-percent weight savings. Achieving a class "A" surface on outer panels and required crash impact response could be barriers to this approach (Brown, 2000; Pryweller, 1999a, 1999b). This technology was also used in the DaimlerChrysler PNGV concept car, the ESX3 (see Chapter 4).

Power Train and Chassis System

The total weight in the power train (868 lbs [394 kg]) and chassis (1,101 lbs [500 kg]) subsystems for the baseline PNGV vehicle is substantial. The total targeted weight savings for the two subsystems is 648 lbs (294 kg), whereas the total identified weight savings, based on the PNGV materials road map, is 325 lbs (148 kg), a shortfall of 323 lbs (147 kg) (NRC, 1999). This shortfall, which was pointed out in the committee's fifth report, has not changed significantly.

Aluminum. Cast aluminum has been used extensively in the power train subsystem, replacing cast iron over the past two decades in most cylinder heads and intake manifolds and increasingly replacing cast iron in cylinder blocks. These trends are expected to continue, although reinforced nylon composites, which provide a cost and weight savings, are replacing aluminum in many intake manifold applications.

The PNGV materials road map describes major efforts to reduce the costs of feedstock and improve the casting processes for cast aluminum (PNGV, 1999a). Progress in the cast light-metals programs includes: creation of a cast light-metal material property database (CD-ROM) that correlates with cast microstructures; the development of a fiber-optic infrared temperature sensor for on-line process control; and the development and validation of a simulation model that predicts monotonic tensile properties of cast metals. This phase of the project is now nearing completion.

Aluminum metal matrix composites applications in chassis and power-train subsystems account for only 30 to 50 lbs. The major hurdles to developing applications of this material are feedstock costs and the development of a reliable

TABLE 2-4 Material Cost Targets

Material	Current Cost ($/lb)	Target Cost ($/lb)	Product Form
Steel	0.30 to 0.40	N/A	Sheet
Aluminum	1.40 to 1.60	1.00	Sheet
Magnesium	1.65	1.20	Ingot
Carbon fiber	> 8.00	3.00	Fiber
Aluminum metal matrix composites	2.00	1.40	Ingot
Titanium	8.00	2.00	Bar, sheet

Source: Adapted from Sherman, 1998.

process for making the composites (Table 2-4). A low-cost powder metal process is under development. Progress includes: the development of materials for a gerotor application; the production of gerotors from production tooling; and initial wear testing.

Magnesium. Projects are under way to improve the machinability of magnesium castings, to develop a lower cost magnesium through the use of plasma torch technology, to develop more creep-resistant alloys, and to utilize semisolid forming (Thixomat, Inc., process) to produce higher strength, low-cost magnesium components. These projects are in the early stages of development.

Titanium. The applications for titanium are in the chassis (40 lb [18 kg] potential savings) and power train (10 lb [4.5 kg] potential savings). The components of interest are springs, piston pins, connecting rods, and engine valves. A program designed to lower the cost of titanium feedstock is in an early stage of development.

Program Assessment and Plans

Body System

Steel/Aluminum Hybrid Body. In the PNGV concept vehicle phase, the emphasis has been on aluminum and composites to meet the 50-percent weight reduction target. As the PNGV program moves toward 2004 and the production of affordable production prototypes, the PNGV team should attempt to balance the opposing requirements of weight reduction and affordability. In the preceding discussion, the committee presented a rationale for considering steel in addition to aluminum and composites (Table 2-4). The PNGV team should carefully follow the progress on the new ULSAB-AVC project. If this modeling study shows the feasibility of a 2,275 lb steel-intensive vehicle, PNGV might be able to add some aluminum and/or magnesium components (hybridize) to the body to produce a 2,000 lb (908 kg) vehicle at a much lower cost than with an aluminum-intensive body.

Aluminum. As a matter of public record, several aluminum-intensive proto-type vehicles have been built outside the PNGV program by the USCAR partners and evaluated for ride, handling, noise, vibration, and handling, crashworthiness, and production processes, such as stamping, extruding, joining, and painting (Jewett, 1997). The three PNGV concept cars are all different in this respect (Askari, 2000). The Ford Prodigy has an aluminum body; GM's Precept has an aluminum skin and frame and composites in the body; the DaimlerChrysler's ESX3 has large integrated plastic body panels supported by an extruded aluminum frame (see Chapter 4). Thus, PNGV partners already have extensive design and manufacturing expertise with aluminum, and a change to an aluminum-intensive vehicle would not be a major technological challenge.

The development of new processing methods, however, would be a challenge especially for feedstock materials, to bring the cost of an aluminum-intensive vehicle to the level of a steel vehicle. As mentioned earlier, the cost penalty of an aluminum BIW over steel is on the order of $1,400 (Schultz, 1999). Current PNGV programs that involve direct casting of thin aluminum sheet, which elimi-nates expensive hot rolling processes, have the potential to reduce the cost of aluminum sheet to $1.00/lb (see Table 2-4).

Injection Molded Composites. The injection molding of large integrated com-ponents from GFRPs being investigated by DaimlerChrysler is conventionally thought to offer weight savings of 25 to 35 percent, which is considerably less than the weight savings with aluminum or CFRP. However, DaimlerChrysler claims a 46-percent weight savings and a 15-percent cost savings over a conven-tional steel body, including the cost of an aluminum space frame to support the composite moldings (DaimlerChrysler, 2000). Through low-cost tooling and part integration, this approach has the potential to be a low-cost solution to body construction.

Carbon Fiber-Reinforced Polymers/Aluminum Hybrid. The committee's fifth report discussed the development of a hybrid-material BIW fabricated largely from thin (1mm) CFRP sheet, which resulted in a weight of 90 kg for the PNGV BIW weighing 285 kg in steel, a 68.5-percent weight savings (NRC, 1999).

Early work by the USCAR Automotive Composites Consortium showed that it was difficult to achieve good frontal crash response from polymer composite front-end structures. Some of these early problems have now been overcome. In the PNGV hybrid-material BIW, lower cost, extruded or hydroformed aluminum beams were used in the front-end structure to ensure adequate crash response, which was verified by 30 mph frontal crash simulations.

The committee considers these results very encouraging, even though there is still much to be learned about: (1) composite design techniques; (2) the consistency of mechanical properties in high production volumes; and (3) the reduction of manufacturing cycle times. Finally, methods of recycling the material into high-value applications to take advantage of its intrinsic properties, as opposed to using it only as filler, must still be developed. But the major barrier to the intensive use of CFRPs for a Goal 3 vehicle is the high cost of the carbon fibers. As shown in Table 2-4, the cost of carbon fiber will have to be reduced from $8.00/lb to $3.00/lb. Currently, the cost penalty of a CFRP BIW is much higher than for a steel or aluminum BIW. Although the weight savings are impressive, CFRP technology must be considered a long-term alternative to steel and aluminum.

Power Train and Chassis System

The committee was pleased to see that several important R&D studies are now in place to reduce the costs and improve the properties of aluminum and magnesium castings. A program is also in place to reduce the cost of titanium feedstock. Cost is the main barrier to increased use of these materials.

Also, the committee notes that programs to help reduce the processing costs of aluminum-metal matrix composite materials have made progress. Gerotors fabricated by the low-cost powder metal-metal matrix composite technology will be wear tested in the coming year. Also, a new project has been initiated on low-cost cast metal matrix composites.

RECOMMENDATIONS

Recommendation. As the PNGV project moves toward 2004 production prototype vehicles, affordability will be a key requirement. Therefore, the development of an efficiently designed and fabricated steel-intensive vehicle being worked on by the American Iron and Steel Institute in the Ultralight Steel Autobody-Advanced Vehicle Concepts (ULSAB-AVC) project should be closely followed, and the possibility of applying the ULSAB concepts to a hybrid steel-aluminum vehicle should be explored.

Recommendation. The committee recognizes the cost reduction potential of DaimlerChrysler's thermoplastic composite injection-molding technology and urges that this work be continued to bring the technology to successful commercialization. The committee encourages the earliest possible generation of vehicle and component test data to define better the structural properties and performance of various composite materials and structures.

Recommendation. The committee regards structural crashworthiness, and safety in general, extremely important in the design of lightweight PNGV vehicles. Using the Oak Ridge National Laboratory car-to-car collision simulation capability, the National Highway Traffic and Safety Administration should support a major study to determine how well lightweight PNGV vehicles would fare in collisions with heavier vehicles and to assess potential improvements.

3

Systems Analysis

PROGRAM STATUS

The systems-analysis model is an important tool for examining the behavior of various overall vehicle system configurations. System models integrate models of individual vehicle components and power train components to predict component and overall system performance. Their primary use is in the development of performance specifications for individual system components to provide optimal overall system performance and in the study of system performance over a range of driving patterns. They can also provide information on future components and total vehicle systems for cost and reliability analysis and for trade-off studies. Over the past few years, the PNGV systems analysis team has developed the PSAT model for these purposes. PSAT is a forward, or "driver-driven," model (i.e., component and vehicle performance are calculated from driver inputs).

During the past year, the systems-analysis team has made encouraging progress in the development of the PSAT model and in providing analysis support to the other PNGV technical teams. A new structure has been set in place for overall development of the PSAT model. Through the national laboratories, DOE has assumed the overall responsibility for funding the development of the PSAT model, in parallel with the development of another vehicle systems model, called ADVISOR, which has been developed by the National Renewable Energy Laboratory. The ADVISOR model is a backward, or drive-cycle-driven, model (i.e., it calculates the power train performance required to drive the vehicle through a specified drive pattern). ADVISOR, which continues to evolve, is available to all users via the Internet; PSAT is available through a secure web site and now has more than 100 registered users. The ADVISOR model runs quickly, requires modest computer resources, and is useful for concept development and evaluation. The PSAT computer code incorporates more sophisticated dynamic models of component behavior and readily allows analysis of transient vehicle performance and control system development. PSAT has been upgraded and improved during the past year in accordance with the PNGV systems-analysis team's work plan, and that process is continuing. An advanced training class for PSAT held in January 1999 was attended by more than 50 people. PSAT has been used to examine the benefits of improved control algorithms (Oakland University) and optimization techniques (University of Michigan) on HEV performance.

The new structure for the coupled development of these two models through joint management and funding by DOE, with industry in a consulting/advisory role, makes excellent sense. Based on different logics, these models are suited to different objectives, and the similarities in many of the subsystem component models will allow developments that improve these subsystem models to be used in both PSAT and ADVISOR.

ASSESSMENT OF THE PROGRAM

The HEV propulsion system, which combines an engine, such as the four-stroke CIDI engine, with energy-storage devices, such as batteries, in a light-weight vehicle, is essential to meeting the 80 mpg fuel economy goal within the time frame of the PNGV program. There are many ways to configure an HEV power train system, and developing and then optimizing such systems with their many interactive components is a challenging task.

The recent announcement by EPA of its proposed Tier 2 emissions standards starting in 2004 has made the modeling of the emissions of HEVs an urgent task for the PNGV systems-analysis team so that extensive emissions/fuel economy/performance/cost trade-off studies can be carried out. Part of this task is to ensure that the emissions models in PSAT and ADVISOR are sufficiently complete and accurate for this purpose. A key issue here is modeling the performance of CIDI engine exhaust catalysts for NO_x and traps for PM and validating these models against experimental data. Because the HEV uses a smaller engine and can control transients through the battery/electric motor propulsion system component, the models should explore whether the HEV configuration provides significant additional opportunities for emissions reduction beyond the lowest emissions levels provided by a stand-alone CIDI engine power train.

A second important task for the PNGV systems analysis team will be to develop a more complete model for fuel-cell power train systems. The fuel-cell-based propulsion system, with its inherently low-emissions characteristics and potential for high efficiency, is a promising longer term technology being pursued by the PNGV program and automobile companies. An unresolved question for the fuel-cell system is how the choice of fuel—hydrogen, methanol, or gasoline/hydrocarbon—will affect the overall system configuration and performance. With hydrogen as the fuel stored on the vehicle, a hybrid system with electrical energy storage will not be required (although it may improve vehicle performance and fuel economy, as well as lower costs). With liquid fuels, the dynamic response of the onboard methanol or gasoline-to-hydrogen reformer is not likely to be adequate to follow vehicle start-up and driving transients, and an HEV system with energy storage will be required. The performance, efficiency, emissions, and costs of these different fuel-cell systems will have to be systematically analyzed. To date, only limited systems-analysis studies of fuel-cell HEV systems have been carried out, and the most promising configurations have yet to be defined. This is an important area for the PNGV systems-analysis team to focus on.

The validation and review of systems models is continuing. The Toyota Prius HEV has been used as a source of data for model assessment. PNGV teams have reviewed critical component models and plans for model improvement. The 4SDI technical team's interaction with the systems model (engine efficiency and emissions predictions, engine mass and warm-up, after-treatment-device models) is especially important. The committee is concerned that PNGV has not devoted enough resources to validation of the PSAT system model. Although validation will be challenging, largely because the data are limited, the PNGV systems analysis team should put more emphasis on this task.

A third important task that should be addressed by the systems analysis team is the development of a generic system/subsystem cost model, which is vital at this phase of the program. The committee recognizes that specific vehicle-level analyses have been performed by the individual automotive companies on a proprietary basis, but a more general model indicating the relative importance of various subsystems and components in the cost structure for different configurations would be of great benefit to sponsors and reviewers of the program. If subsystem cost targets have not been met, trade-offs may be necessary, perhaps even in the goals of the program.

Other sections of this report highlight the need for systems analysis of trade-offs among performance, fuel economy, emissions, and cost for alternative primary power plants, fuel composition, degree of hybridization (relative magnitude of engine and battery/motor power, use of regenerative braking for recharging), and impact of accessory loads. For fuel cells, important issues are the trade-offs between efficiency of the fuel cell, overall system energy conversion efficiency, component sizes, and cost. For batteries, an important question is how failure to attain cost goals will influence the optimum degree of hybridization and the overall system energy conversion efficiency.

Recommendation. Given the potential of fuel-cell technology for meeting the efficiency and emissions objectives of the PNGV program, the systems-analysis team should increase its efforts to develop more complete and accurate fuel-cell system and component models to support the development and assessment of fuel-cell technology.

4

Concept Vehicles

The PNGV plan calls for the construction, development, and testing of concept vehicles during 2000. These concept vehicles are intended to establish the functional benefits of a design but may use components for which validated manufacturing processes and affordable costs have not yet been established. All three manufacturers are well into this stage of the program. Ford and GM introduced cars at the North American International Automobile Show in Detroit in January 2000, and DaimlerChrysler introduced its car to the public at the National Building Museum in Washington, D.C., in February 2000. As expected, each manufacturer has taken a somewhat different approach, but the concept cars all share technology and know-how developed by PNGV (some of which is finding its way into current production vehicles as called for in Goal 2). Table 4-1 provides a comparison of the attributes of the PNGV concept vehicles; Table 4-2 provides a comparison of the Toyota and Honda hybrid vehicles. Figures 4-1, 4-2, and 4-3 are photographs of the three PNGV concept vehicles.

All of the concept cars incorporate hybrid-electric drive trains designed around small turbocharged CIDI engines that shut down when the vehicle comes to rest. They are all based on sophisticated structural optimization techniques and high strength-to-weight materials, such as aluminum and composites in both bodies and interiors. The design of every aspect of these cars, including wheels, tires, interior components, front, back, and side windows, rear vision devices, and aerodynamic drag, has been modified to reduce weight and increase efficiency. Friction has been reduced in almost every rotating component. These cars are expected to achieve 70 to 80 mpg (gasoline equivalent), although, to date no tests have been run to confirm these figures. Emissions are targeted to meet Tier 2 standards, but the after-treatment systems necessary to achieve these levels have not yet been defined or validated. In summary, concept cars represent a major milestone toward meeting PNGV Goal 3, and each contributes significantly to our understanding of the challenges this goal represents. The committee congratulates the USCAR partners for their outstanding efforts and the demonstrated results.

GENERAL MOTORS

The GM concept vehicle, the Precept, is designed to meet all of the Goal 3 functional targets (with the exception of luggage space) and is expected to achieve 80 mpg using a dual-axle regenerative parallel hybrid power train (GM, 2000). In addition to this design, GM has built a fuel-cell version of the Precept packaged in the same basic chassis and body. The fuel-cell vehicle uses a hydride hydrogen storage system and is expected to achieve a gasoline equivalent fuel economy of more than 100 mpg and have a range of 500 miles without refueling. A fully functional version of this car is expected by the end of 2000.

The Precept hybrid is a very complex vehicle and represents state-of-the-art performance in all of its design parameters and components. The body structure is an aluminum-alloy space frame with both aluminum and plastic exterior panels. Every component has been designed to fulfill its function with minimum weight. The aerodynamic drag coefficient has been reduced to 0.163, the lowest known value for a five-passenger automobile and lower than that of any production vehicle. The propulsion system consists of a 1.3-liter, three-cylinder, CIDI, turbocharged, 44-kW aluminum engine coupled with a 10-kW peak power electric motor and alternator mounted in the rear of the car, plus a 25-kW peak power electric motor and alternator in the front. The rear unit employs a four-speed, automatically shifted, manual transaxle. This configuration permits regenerative braking on all four wheels, the energy is stored in either a lithium-polymer or a NiMH battery pack, with a 3-kWh usable energy capacity. Curb weight of the vehicle is 2,590 lbs (1,176 kg); the wheel base is 112 in (2,839 mm); the overall length is 193 in (4,906 mm); and overall width is 68 in (1,726 mm).

The sophistication evident in the overall design and electronic controls for energy management, thermal management, chassis systems, and climate control for the Precept hybrid show how challenging meeting the PNGV Goal 3 is. This vehicle demonstrates what can be achieved with today's technology but raises significant questions about the affordability of meeting the 80-mpg fuel economy target. Since the inception of the PNGV program, significant progress has been made in reducing the cost of many components, especially the power electronics, but a very significant cost challenge still remains.

The Precept fuel-cell vehicle is designed with 105-kW peak, 75-kW continuous power, hydrogen-air PEM fuel cell operating at 1.5 bar, together with two hydride storage tanks expected to hold a total of 4.9 kg of hydrogen, all within the vehicle structure described above. In addition to the fuel-cell stack, the system

TABLE 4-1 Comparative Attributes of PNGV Concept Vehicles

Attributes	PNGV Targets	DaimlerChrysler Dodge ESX 3	GM Precept Hybrid	FORD P2000 Prodigy Hybrid
Acceleration	0–60 mph in 12 sec	0–60 mph in 11 sec	0–60 mph in 11.5 sec	0–60 mph in 12 sec
Passenger capacity	up to 6	5	5	5
Fuel economy	up to 80 mpg	72 mpg	80 mpg	>70 mpg
Range	380 miles	400 miles	380 miles	> 660 miles on diesel fuel
Emissions	meets standards	targeted for Tier 2	Tier 2 at default levels[a]	targeted for Tier 2
Luggage capacity	16.8 ft³, 200 lbs	16 ft³	4.4 ft³	14.6 ft³
Recyclability	80%	80%	N/A	
Safety	meets or exceeds FMVSS	meets FMVSS	meets FMVSS	meets FMVSS
Cost/affordability	equivalent to current vehicles	$7,500 price premium	N/A	not affordable[a]
Length		4,902 mm (193.0 in)	4,906 mm (193.2 in)	4,747 mm (186.9 in)
Width		1,882 mm (74.1 in)	1,726 mm (67.9 in)	1,755 mm (69.1 in)
Height		1,402 mm (55.2 in)	1,383 mm (54.4in)	1,419 mm (55.9 in)
Body structure		LIMBT on aluminum frame	aluminum alloy space frame	aluminum unibody construction
Curb weight	898 kg (1,980 lbs)	1,021 kg (2,250 lbs)	1,176 kg (2,590 lbs)	1,083 kg (2,385 lbs)
Aerodynamics (C_d)	0.2	0.22	0.163	0.199
Heat engine Type		DDC (CIDI) 3 cylinder / 12 valve / turbo (VNT)	ISUZU (CIDI) 3 cylinder / 12 valve / turbo (VNT)	DIATA (CIDI) 4 cylinder / 16 valve / intercooled turbo
Displacement		1.47 liters	1.3 liters	1.2 liters
Power output		55 kW (74 hp) @ 4,200 rpm	44 kW (59 hp) @ 3,000 rpm[b]	55 kW (74 hp) @ 4,500 rpm
Torque		165 Nm (122 ft-lbs) @ 2,200 rpm	170 Nm @ 2,000 rpm	153 Nm (113 ft-lbs) @ 2,250 rpm
Fuel tank capacity		6 gallons (diesel)	4.5 gallons (CA low-sulfur diesel)	(diesel)

Electric motor	DC brushless PM 110-195 V DC	Panasonic (350 V 3 Phase AC)[c]	350 V 3 phase induction
Maximum power	15 kW (20hp) peak power/ 3 kW continuous	25 kW (34 kW) peak power/ (16 kW continuous)	35 kW (47 hp) peak power/ (3 kW continuous)
Batteries	Li-ion 150 V peak power 22 kW (28.2 hp)	NiMH 350V, 3 kW-hr usable 28-12V modules with liquid thermal mgmt.	NiMH 288V, 1.1 kW-hr
Transmission	EMAT - six speed	automatic shifted manual	5-speed automated manual
Braking	front axle regenerative braking	4-wheel regenerative braking	front wheel regenerative braking

[a] Tier 2 default standards were Tier 2 emission levels in the Clean Air Act for consideration by EPA. Since the announcement of Tier 2 standards, these are no longer relevant.

[b] A 10-kw electric motor/generator attached to the engine supplied by the battery adds to the propulsive power during maximum acceleration.

[c] GM has added a rear electric motor from Unique Mobility that contributes to kW of peak propulsion power and 150 Nm (110 ft-lbs) of torque.

[d] Ford has committed to production of an HEV with a $3,000 cost premium (assumed to use gasoline).

Acronyms: CVT = continuously variable transmission; DIATA = direct-injection, aluminum-block, through-bolt assembly; DOHC = dual overhead cam; EMAT = electro-mechanical automatic transmission; FMVSS = federal motor vehicle safety standards; LIMBT = light weight injection molded body technology; MTX = manual transaxle; SULEV = super ultra low emission vehicle; ULEV = ultra low emission vehicle; VNT = variable nozzle turbo; VTEC = variable valve timing and lift electronic control.

Source: Adapted from USCAR.

TABLE 4-2 Comparison of the Toyota and Honda Hybrid Vehicles

	PNGV Targets	Toyota Prius (Japanese Specifications)	Honda Insight
Acceleration	0–60 mph in 12 seconds	0–60 mph in 14.1 seconds	0–60 mph in 12 seconds
Passenger capacity	up to 6	5	2
Fuel economy	up to 80 mpg	56 mpg[a]	64 mpg
Range	400 miles	550 miles	600 miles
Emissions	meets standards	SULEV	ULEV
Luggage capacity	16.8 ft³, 200 lbs	10 ft³	5.5 + 1.5 ft³
Recyclability	80%		
Safety	meets or exceeds FMVSS		
Cost affordability	equivalent to current vehicles	$17,500 cost premium	$10,000 cost premium
Length		4,274 mm (168.3 in)	3,939 mm (155.1 in)
Width		1,694 mm (66.7 in)	1,694 mm (66.7 in)
Height		1,490 mm (58.7 in)	1,353 mm (53.3 in)
Body structure		conventional unibody (steel)	aluminum unibody
Curb weight	898 kg (1,980 lbs)	1,240 kg (2,734 lbs)	835 kg (1,856 lbs)
Aerodynamics (C_d)	0.2	0.30	0.25
Engine heat		Gasoline	gasoline
Type		inline 4 cylinder DOHC	3 cylinder /12 valve / VTEC
Displacement		1.5 liter	1.0 liter
Output (hp)		43 kW (58 hp) @ 4,000 rpm[b]	50 kW (67 hp) @ 5700 rpm
Torque		102 Nm (75 ft-lbs) @ 4,000 rpm	90 Nm (66 ft-lbs) @ 4,800 rpm
Fuel tank capacity		13.2 gallons (gasoline)	10.6 gallons (gasoline)
Electric motor		permanent magnet (274 V)	"ultra-thin motor/gen" (144 V)
Maximum power		33 kW (44 hp)	10 kW (13.5 hp)
Batteries		Panasonic 288 V NiMH 240 D-size modules	Panasonic 144 V / 65 amp-hr 120 D-size modules
Transmission		electronically controlled CVT	5 speed MTX
Braking		front wheel regenerative braking	front wheel regenerative braking

[a] Second generation Prius, expected to be for sale in the United States in June 2000.

[b] Engine is revolution limited to 4,000 rpm to allow engine components to be built lighter (Atkinson Cycle).

Acronyms: CVT = continuously variable transmission; DOHC = dual overhead cam; FMVSS = federal motor vehicle safety standards; MTX = manual transaxle; SULEV = super ultra low emission vehicle; ULEV = ultra low emission vehicle; VTEC = variable valve timing and lift electronic control.

Source: Adapted from USCAR.

FIGURE 4-1 The General Motors Precept concept vehicle. Source: GM and USCAR.

requires an air compressor, a humidifier, a heater for releasing the hydrogen, a coolant pump, and several heat exchangers. Packaging all of these components without infringing on vehicle utility presents a major challenge. The Precept fuel-cell vehicle shown to the public demonstrates that this packaging can be accomplished. An operating version of this vehicle is anticipated by the end of 2000.

FORD

The Ford Prodigy, also an HEV concept car, is designed to meet all of the functional requirements of Goal 3 and deliver about 70 mpg (Ford, 2000). Ford has chosen a "low storage requirement hybrid design" for this car. The power train, which is considerably simpler than the power train for the GM Precept, consists of a 1.2-liter, four-cylinder, CIDI, turbocharged 55-kW aluminum engine with a starter/alternator that replaces the engine flywheel. A five-speed, automatically shifted, manual transmission is used. The starter/alternator is rated at 3 kW continuous, 8 kW for three minutes, and 35 kW for three seconds. The hybrid NiMH battery is designed for high power, but its storage capacity is only 1.1 kWh. The Prodigy vehicle has an all aluminum body with a wheel base of 109 in (2,781 mm), an overall length of 187 in (4,747 mm), an overall width of

FIGURE 4-2 The Ford Prodigy concept vehicle. Source: Ford Motor Company and USCAR.

69 in (1,755 mm), a curb weight of 2,385 lbs (1,083 kg), and an aerodynamic drag coefficient of 0.199.

Although there are many similarities between the Ford Prodigy and the GM Precept, they represent quite different approaches to meeting Goal 3. Because the Ford power train system is considerably simpler, it is closer to meeting the affordability target of Goal 3, although it sacrifices some fuel economy by limiting the amount of potential regenerative braking. These kinds of trade-offs will have to be made as manufacturers attempt to determine the market acceptability of any of these concepts.

Ford has also demonstrated a fuel cell-powered version of the P-2000 car, predecessor of the Prodigy. This vehicle stores gaseous hydrogen under high pressure.

DAIMLERCHRYSLER

The power train in the DaimlerChrysler ESX3 PNGV concept car is very similar to the one in the Ford Prodigy (DaimlerChrysler, 2000). The battery, however, is expected to be a Li-ion pack with a peak power capability of 22 kW

FIGURE 4-3 The DaimlerChrysler ESX3 concept vehicle. Source: DaimlerChrysler Corporation and USCAR.

to supply a DC brushless, permanent-magnet motor instead of a three-phase induction machine. Its six-speed manual transmission has two clutches to smooth shift transitions.

The most innovative feature of the DaimlerChrysler ESX3 is its large injection-molded plastic body sections, which presents the possibility of building a body structure that is simultaneously both lighter in weight and lower in cost than a conventional steel body and fully recyclable. The curb weight of the ESX3, 2,250 lbs (1,021 kg), is lower than the curb weights of the other two concept vehicles(Prodigy: 2,385 lbs (1,083 kg); Precept: 2,590 lbs (1,176 kg). The aluminum body construction used by both the Ford Prodigy and the GM Precept cars is currently significantly higher in cost than a comparable steel body. DaimlerChrysler estimates that the ESX3 could have a $7,500 price premium (Robinson, 2000). The other companies did not provide price estimates.

DaimlerChrysler is pursuing hybrid electric propulsion technology using a CIDI engine in much the same way as the other two manufacturers. Nevertheless, even before the merger with Chrysler, Daimler had shown great interest in fuel cells and had demonstrated an A-class hydrogen-powered fuel-cell car that stores hydrogen as a liquid.

SUMMARY

The construction and public demonstration of these concept vehicles are the result of massive efforts by all of the car companies and is a major step forward for the PNGV program. USCAR solicited its members for expenditures on "PNGV-related" research and estimated the combined annual expenditure of the three companies for the last three years was about $980 million. As development on these vehicles proceeds, much valuable information will be obtained to guide the extensive performance improvements, cost reductions, and manufacturing development that will be necessary to move to the next stage of the PNGV program, production-ready prototypes by 2004.

5

Major Crosscutting Issues

This chapter first briefly reviews some advantages of a technology partnership, such as the PNGV program, and provides definitions of program success. The remainder of the chapter examines the major achievements and barriers, the adequacy and balance of the program and potential future directions, issues raised by the Tier 2 emission standards, and timely consideration of fuels issues for new automotive power plants.

BACKGROUND

The PNGV Concept

The basic concept underlying the automotive industry/government partnership is the fulfillment of automobile-related societal goals, as perceived by the federal government, with minimum disruption to the industry's ability to meet marketplace demands. The PNGV was formed to develop technology through cooperative research among the three automobile companies and federal research entities that would enable a substantial improvement in the fuel economy of new vehicles without sacrificing desirable market characteristics. If the PNGV program is successful, it would avoid the conflicts inherent in "technology-forcing" regulations, which run the following risks:

- requiring that technologies be introduced before predictable field performance and reliability have been established, possibly causing undesirable, and sometimes unexpected, consequences

- increasing costs, causing a decrease in new car sales, thereby limiting the rate of improvement in the societal benefits of new technologies
- causing significant undesirable economic effects on the industry and the country

A partnership for technology development has several advantages. The resources of both the industry and a variety of government laboratories and universities provide a broad base for research and create a shared understanding of the solutions and trade-off possibilities. The lessons learned from the program could be used to inform future public policy decisions that may be less disruptive to industry facing regulatory requirements. Ultimately, this approach might be extended to enable a more rapid deployment of effective solutions to a broad range of societal problems, and, at the same time, promote a better understanding of potential side effects of changes.

Definitions of Program Success

PNGV's approach has been to set one "stretch" goal with specific criteria and a 10-year deadline for "production-ready" technology together with a "best-efforts" fuel economy target of 80 mpg (gasoline equivalent). The program also set two other goals to encourage the near-term application of research results. Success, therefore, can be defined in many ways. The committee used the following criteria for determining success:

1. the attainment of all aspects of the "stretch" Goal 3, namely, the development by 2004 of a production prototype midsize sedan that meets all emission and safety standards, has a fuel economy up to 80 mpg, and costs no more than conventional 1994 family sedans, adjusted for economics
2. the development of vehicles by 2004 with a fuel economy and cost trade-off that maximize potential market penetration and meet Tier 2 emission requirements
3. the transition of as much of the technology developed as possible to a wide range of production vehicles (Goal 2) with significant progress toward a state of technology beyond 2004 that will be much more fuel efficient

In the committee's view, the simultaneous attainment of the three critical requirements (emissions, fuel economy, and cost) of Goal 3 by 2004 is very unlikely. Although the 80-mpg fuel economy level appears to be technically feasible, the cost requirement is clearly unattainable with known or projected technological development in the program schedule (2004). Also, it appears that meeting the Tier 2 emissions standard will result in a fuel economy well below

80 mpg, and, even then, it will be difficult to achieve in production vehicles with adequate probability for meeting the certification period of 100,000 miles. The development of vehicles of radical design (e.g., a fuel-cell vehicle) for mass production by 2004 is also highly optimistic.

The second definition of success, although it does not include the cost parity envisioned in the original goal, recognizes that the ultimate objective of reducing fuel consumption would be served by achieving large market penetration of the new technologies. In effect, 60 mpg instead of 80 mpg would still represent a major reduction in fuel consumption. For a car traveling 15,000 miles per year, the baseline vehicle would use 560 gallons, the 60-mpg car 310 gallons, and the 80-mpg car 190 gallons.

In the third definition, success is reflected by the commercial introduction of radically new technology, such as a fuel-cell power plant, rather than the construction of a specific short-term prototype production vehicle.

MAJOR ACHIEVEMENTS AND TECHNICAL BARRIERS

Goals 1 and 2

Although most of the discussion about achievements and barriers is directed toward Goal 3, the committee found evidence of continuing and significant progress toward achieving goals 1 and 2:

- the successful completion of a project to demonstrate continuous cast sheets of Series 5000 aluminum for body structures and a follow-up project to develop similar processes for exterior body parts
- several smaller efforts to expand aluminum manufacturing and assembly capabilities and an alliance between the automotive and aluminum industries to address standardization, scrap recovery, and other issues
- cost reduction of carbon-fiber composites, improvement of their properties, and development of new manufacturing techniques, as well as the recycling and design of hybrid material bodies
- the development of techniques for predicting aluminum springback

Goal 3 Achievements

Substantial technical progress has been made in reducing the energy required to propel the vehicle (e.g., reduced mass, drag, etc.) and supplying auxiliary loads (e.g., heating, air conditioning, etc.). Simultaneous efforts have resulted in continued improvements in the efficiency and performance of power plants (e.g., 4SDI engines, fuel cells), performance and life of energy storage devices (batteries), and in the development of modeling and simulation techniques. The three concept vehicles described in Chapter 4 show the results to date of these substantial efforts. Major achievements for specific components are detailed below.

Vehicle Engineering

A number of accomplishments have been achieved in vehicle engineering, including the following:

- the fabrication and testing of a lightweight hybrid material body to validate weight reduction of more than 40 percent
- the completion of an energy-efficient occupant-comfort project with a 75 percent reduction in required energy achieved, for example, by reduced thermal mass of the vehicle interior, improved efficiency of the heating and cooling systems, and optimized thermal management
- the completion of a lightweight interior project demonstrating a 157-lb (71-kg) interior weight reduction
- initiation or continuation of projects to address issues for a high (42 percent) payload/curb weight ratio, low rolling resistance (run-flat) tire, underbody airflow management, and energy-efficient side window development

Engines and Fuels

The following accomplishments have been achieved in the engines and fuels areas:

- New collaborative projects have been initiated in advanced combustion and emission controls (e.g., Detroit Diesel and Johnson-Matthey; Cummins Engine Company and Engelhard). Other continuing projects are advancing the understanding of catalysts, as well as defining fuels issues.
- SNL has developed a new catalyst with lower "light-off" temperature and better NO_x reduction. SNL is also pursuing a novel means for reducing PM emissions, improving the effectiveness of EGR, and gaining a better understanding of combustion processes.
- LANL has developed a new zeolite-supported catalyst to improve NO_x reduction and has formed promising microporous catalysts.
- A project by Industrial Ceramic Solutions has resulted in a PM filter that can remove up to 90 percent of diesel particulates and can be regenerated at idle using microwave techniques.
- Southwest Research Institute has demonstrated that fuel formulation could reduce diesel PM emissions by 50 percent and NO_x by 10 percent. ORNL is using its refinery models to evaluate the impact of various formulations on the cost of diesel fuel.

- PNNL has developed and tested a plasma catalyst that shows high conversion rates even in the presence of sulfur.

Batteries

Accomplishments in the battery area include the following:

- receipt and evaluation of a 50-V NiMH module from VARTA
- receipt of four Li-ion modules from SAFT for testing
- the initiation of a project to build a 300-V system at SAFT
- the identification of a Li-ion electrochemistry projected to increase calendar life from two years to three to five years
- better understanding of failure mechanisms and abuse tolerance issues
- the incorporation of NiMH batteries in the Prodigy concept vehicle
- the incorporation of Li-ion batteries in the ESX3 concept vehicle
- the incorporation of NiMH (and later lithium-polymer) batteries in the Precept concept vehicle

Power Electronics

The power electronics and electrical systems area has made progress in the following areas:

- SNL is improving DC high-voltage-bus capacitors. Results to date indicate that improved performance and reduced cost are feasible.
- A project at ORNL is helping to assess the mechanical reliability of electronic ceramic devices and to identify less expensive alternatives through mechanical characterization.
- ANL and ORNL are working to develop processes to fabricate neodymium-iron permanent magnets with up to 25 percent higher magnetic strengths than with available magnets. A facility has been completed and characterization has begun.
- ORNL has developed a 100-kW inverter with a power density of 11 kW/kg and is working with ORNL, Wright Patterson Air Force Base, the Electric Power Research Institute, the U.S. Department of Defense, and SNL on a 100-kW, reduced cost, motor controller.

Fuel Cells

The following significant fuel-cell developments have been made:

- the operation of an Epyx gasoline (fuel-flexible) fuel processor in conjunction with a 10-kW Plug Power PEM stack

- the demonstration of microchannel fuel processing of iso-octane (PNNL)
- the demonstration of a new higher temperature nonair-sensitive fuel reformer catalyst (ANL)
- the demonstration of a much more CO-tolerant anode (LANL)
- industry demonstrations of a high power density stack (AlliedSignal), low-cost composite bipolar plates (Institute for Gas Technology), and low-cost membrane electrode assemblies (3M)
- continued improvements in modeling and simulation (ANL, LANL)
- the demonstration of production techniques for low-cost molded bipolar plates (LANL, Premix, Inc.)

Goal 3 Barriers

In spite of substantial accomplishments in virtually every area of PNGV activities, formidable barriers remain in virtually every area.

- New business arrangements, such as the Daimler-Chrysler merger and the Delphi spin-off from GM, as well as the fact that the program has moved into the prototype-development phase, have made reaching consensus on precompetitive projects more difficult.
- Neither aluminum nor composite materials are yet projected to reach costs competitive with steel for most major vehicle components. Nevertheless, aluminum and composite materials are essential to meeting the weight reduction targets.
- To meet projected emission requirements, CIDI engines will require a new fuel (e.g., with low sulfur and aromatics), a low-cost after-treatment system with NO_x reduction efficiencies of more than 75 percent, at least a 50-percent effective PM trap, and a minimal effect on fuel economy.
- New sensor and control technology will be necessary, as well as cost reductions for common-rail fuel injection, for all 4SDI engines.
- For advanced CIDI engines, spark-ignition, direct-injection engines, and gasoline-fueled fuel-cell systems, either a low-sulfur fuel (e.g., <10 ppm) or an onboard method of removing sulfur will be necessary to avoid the deactivation of catalysts. EPA recently proposed a diesel fuel sulfur control requirement of no greater than 15 ppm beginning June 1, 2006 (EPA, 2000).
- Costs of batteries are still projected to be at least three times targeted costs. Both life and abuse tolerance issues must also be resolved.
- System complexity and operation under all ambient conditions are major problems for gasoline (fuel-flexible) fuel-cell systems, and projected costs are still about six times too high for a cost-competitive power plant.

- The size and weight of fuel-cell fuel processors are still much too high.
- No liquid fuel-cell system tests, or even projections of known technologies, indicate start-up times of less than several minutes, which will not be acceptable to consumers.
- Better oxygen reduction catalysts and higher CO tolerance will be necessary to move towards fuel-cell cost and performance targets.
- An efficient, lightweight, low-cost, quiet, fuel cell compressor/expander has not yet been designed.
- High-volume, low-cost manufacturing techniques will have to be developed for much of the power electronics.
- Integrated thermal management for power electronics, which will be necessary for efficiency, life, size, cost, and performance, is still complex and costly.
- Fuel supply infrastructures for fuels other than gasoline have yet to be put in place.

ADEQUACY AND BALANCE OF THE PNGV PROGRAM

Distribution of Funding

Figure 5-1 shows an industry analysis of funding by technology for PNGV from the DOE Office of Advanced Automotive Technologies over the life of the PNGV. The decrease in funding for hybrid propulsion systems since 1998 is substantial, as is the corresponding increase in funding for fuel cells. The decrease in hybrid systems was apparently caused by the conclusion of the DOE contract for the Hybrid Propulsion System Development Program and was more than compensated for by the increase in industry effort on hybrid systems for the 2000 concept cars. The increase in funding for fuel cells was apparently the result of optimism about their eventual success. The committee believes government funding for longer range, precompetitive research and industry efforts focused more on near-term development are entirely appropriate.

Figure 5-1 shows that total DOE funding in 2000 is expected to be approximately $128 million. This, of course, is not the entire funding for PNGV, and DOE's allocation among technologies does not represent the overall distribution of effort among technologies for the PNGV. About another $110 million of PNGV-related funding is expected to be provided by other government agencies (the U.S. Department of Commerce, EPA, U.S. Department of Transportation, National Science Foundation), perhaps half of which will go to emissions control and half to long-range research. The committee was informed that USCAR solicited its members, on a confidential basis, for overall figures on expenditures for "PNGV-related research" and arrived at a total investment for 1999 of

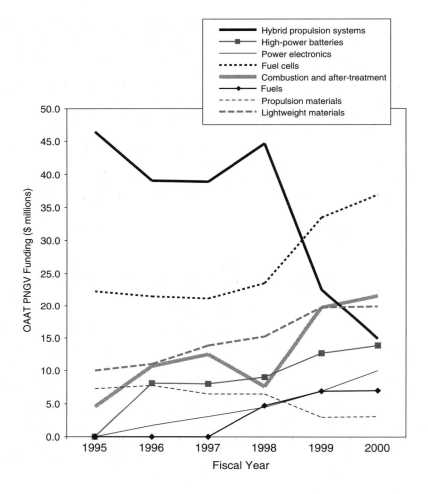

FIGURE 5-1 Distribution of DOE's Office of Advanced Automotive Technologies budget for PNGV (by technology). Source: PNGV, 1999b.

$982 million. The three USCAR companies estimated that their investments for each of the previous three years of the program was approximately the same. This very large investment (far above the program's expected 50/50 government/industry matching level) represents a major effort on the part of the industry partners to develop the concept cars for 2000.

Criteria for Adequacy and Balance

The allocation of resources (e.g., money, people, and facilities) to the PNGV program should optimize the chance of success of the program. Critical criteria for judging the adequacy and balance of the program are listed below:

- Are some important projects receiving too few resources?
- Are some projects receiving too many resources or duplicating other efforts?
- In the face of limited budgets, should some projects be reduced in favor of higher priority projects?
- Should some projects with poor chances of success be eliminated?

In short, major efforts should be devoted to solving major problems, which are listed below:

1. *Fuel economy and emissions trade-offs.* Radical after-treatment technology will reduce fuel economy of the CIDI engine, and the substitution of a gasoline spark-ignition engine to make it easier to attain the emissions and cost goals will still compromise fuel economy. However, the gasoline engine uses technology that is more developed and much closer to production readiness. A comparison of the fuel economy outcome of these two engine applications would be very useful.
2. *Meeting Tier 2 emission requirements with a CIDI engine.* After-treatment systems for control of NO_x and PM emissions from CIDI engines are under intensive study, but the proposed systems appear to be cumbersome, expensive, and not effective enough to meet Tier 2 standards. The possible emission benefits of an HEV configuration with electric motors and batteries have not yet been determined and should be an immediate subject of study in the PNGV. Optimizing the system for emissions rather than for fuel economy should also be given immediate consideration.
3. *High costs.* Cost is a serious problem in almost every area of the PNGV program. In fact, DaimlerChrysler announced that the selling price of its concept HEV, the ESX3, if put into production, would be $7,500 more than for a conventional car. Costs of most components in the proposed vehicle systems are above the target values, and rudimentary estimates of the cost penalties for complete vehicle systems are several thousands of dollars.
4. *Systems analysis.* Reasonable systems analysis tools have now been developed, but, as yet, the PNGV has not applied them to the many outstanding questions of HEV design and performance. Both emissions and costs remain to be modeled. Systems analysis should be used for

trade-off studies, which will be critical to decisions on program direction
and the actual design of systems.

5. *Battery and power electronics cost and performance.* Both of these are
far from their targets, which could well be a barrier to the realization of a
market-acceptable HEV. Technical breakthroughs or strategies to overcome this barrier will be necessary.

Figure 5-1 shows that the major DOE funding is being devoted to some of
these major problems, namely, emissions after-treatment, high-power batteries,
and power electronics. The development of HEV propulsion systems is the object
of major efforts by industry, and the committee was reassured to see that serious
concerns about costs were raised in all of the presentations at the committee
meetings. It was not clear from the estimates provided to the committee whether
or not systems analysis is being appropriately funded, but more effort in this area
is clearly needed.

Long-Range vs. Short-Range Research and Development

If one adopts the first definition of success (meeting all aspects of Goal 3
with a production prototype by 2004), then the PNGV has very little time and
faces immense challenges. To satisfy this definition of success, PNGV would
have to focus more on the major problems listed above, which probably would
require cutbacks in other longer-range projects, assuming funding did not
increase.

If one adopts the second definition of success (fuel economy/cost trade-off to
maximize market penetration and meet Tier 2 standards), then development can
proceed along more orderly lines with a broader program, a better chance of
achieving breakthroughs, and a better chance of having a marketable product
with reliable performance. To satisfy this definition of success (which is favored
by the committee), the federal administration and Congress (if they want to
promote the early deployment of the high fuel economy PNGV-type vehicles)
may have to evaluate the advisability of temporary incentives (e.g., tax rebates) to
offset the higher initial vehicle costs.

Adopting the third definition of success (to pursue Goal 2 aggressively and
work toward fuel efficiency beyond 2004) would leave time to solve the major
problems listed above and would allow for a more deliberate allocation of
resources for Goal 2 and a drastic reduction in funding, or perhaps the abandon-
ment of projects that have limited chances of success. Fuel-cell vehicle research
could be the major focus of continuing efforts because fuel-cell vehicles appear
to be far from production-ready at this time.

The committee encourages the PNGV leadership to develop specific objectives
for the production-prototype phase of the program with the following objectives
in mind. First, each automotive company should develop production-feasible

vehicles that come as close as is practical to the original vehicle performance objectives of Goal 3, meet the mandated emission requirements, and balance the inevitable shortfalls in fuel economy, vehicle performance, and affordability to maximize potential market acceptability. Second, production-feasible versions of new PNGV component technologies that can, in an evolutionary way, be incorporated into new vehicle designs under Goal 2, should be developed. The first objective will continue to "stretch" the new technologies and system concepts that have the potential to provide large improvements in fuel economy in the vehicle fleet. The second objective will prompt the development and application of component technologies critical to improving fuel economy.

In the committee's judgment, the Tier 2 NO_x and PM emission standards as currently promulgated could potentially exclude the CIDI internal combustion engine, with its significant fuel economy benefit, from early introduction in the United States. Even if it is not excluded, measures to meet the Tier 2 standards may reduce the CIDI engine's fuel economy to such an extent that it is no longer more efficient than alternative engines. Therefore, the new emissions requirements may require PNGV to shift its development efforts away from the highly efficient CIDI engine and toward the adaptation of other internal combustion engines that have more potential for extremely low emissions. USCAR and the government agencies involved in the PNGV should begin serious discussions about whether lesser improvements in fuel economy with the alternative engines available for the next phase of the program (the 4SDI spark-ignition engine and the port fuel-injected gasoline spark-ignition engine), which have significantly better potential for meeting the Tier 2 emission levels, is an appropriate trade-off from a national perspective. A wiser choice might be to extend the deadline for meeting the fuel economy target and lower emissions objectives and allow more time for the development of new fuel economy technology. PNGV will have to clarify the objectives of the production prototype phase of the program, especially in light of the changed emission standards.

Constraints

Budgeted federal expenditures are an obvious constraint to PNGV's overall efforts, as well as to some individual projects. PNGV management does not have complete control over these amounts. By the same token, they have little control over the expenditures of industry in support of PNGV. In 1999, the industry clearly put a tremendous effort into developing and producing the concept vehicles, which have addressed nearly all of the major problems and made a giant leap toward satisfying the PNGV goals and defining the remaining challenges.

Adequacy of Resources

The adequacy of funding of PNGV is difficult to assess because the funding figures provided to the committee are incomplete. It could be said that, because progress toward Goal 3 appears to be insufficient to attain the objectives by 2004, the resources are inadequate. However, the industry partners have stated that the greatest limitation they face is the lack of talented people rather than the lack of money and that new ideas (breakthroughs) are needed more than dollars. The committee is inclined to agree, although increased funding might accelerate some projects and support a broader program with more likelihood of breakthroughs. However, the committee could find no specific areas in the PNGV program that are starving for funds. Therefore, the committee concluded that the overall supply of resources is appropriate.

Balance of Resources

The balance of resources in PNGV is even more difficult for the committee to assess than funding because no data are available on the industry distribution of funding by project. The balance shown in Figure 5-1 appears to be appropriate based on the past history of PNGV, if government funding is supposed to be applied mainly to long-range R&D. The committee assumed that, in the course of their development of the concept vehicles, industry resources were directed heavily toward solving the major problems of HEV optimization. However, looking ahead to 2004, a different balance of effort will be required, and the new balance will depend heavily on the course each company chooses to follow. For example, if the companies choose to continue working toward production prototypes with a CIDI engine in an HEV configuration, then a major effort will have to be mounted on emissions control for that power plant and a determination made of the benefit of optimizing that system for emissions control rather than for efficiency. This determination will also determine the fuel economy and cost penalties of meeting the Tier 2 emission standards with whatever technology can be developed. If the companies choose to replace the diesel engine with a gasoline spark-ignition engine, which they have said can meet the Tier 2 requirements, then the optimized fuel economy of that configuration can be compared to the fuel economy of the diesel system. The gasoline system will probably have a somewhat smaller cost penalty than the diesel configuration and reach production readiness more quickly. The committee believes that both of these options should be investigated using the best systems analysis and experimental evidence available, and the balance of future R&D should be adjusted according to the results.

Cost obviously remains a serious problem in almost all areas of PNGV and will have to be addressed between now and 2004. Although the problem of cost has received more attention in the past year by the PNGV technical teams, no major changes in resource allocation have been made as a result.

The federally funded longer term research shown in Figure 5-1 may well be continued in anticipation that breakthroughs useful to the industry will be made. However, industry representatives have complained that much of this research is not likely to be helpful in the near term. Better communication between the federally funded researchers and industry engineers might correct this problem.

With no breakthroughs, the committee now believes that the likely outcome for 2004 will be an HEV system that approaches the Tier 2 emission standards but has a fuel economy somewhat lower than that of the concept cars and a cost several thousand dollars higher than conventional vehicles. Based on the data presented to the committee, the fuel-cell vehicle program (and its required infrastructure) appears to be on a separate, longer range schedule that extends beyond the 2004 production-prototype development time frame. The federal resources for fuel-cell vehicles seem to be appropriate for this longer range technology, and industry appears to be devoting substantial resources to fuel cells. Because work on the fuel-cell vehicles has made considerable progress and is still considered promising for beyond the 2004 milestone, the committee sees no reason for reallocating those resources.

Recommendation

Recommendation. At this stage of the program, PNGV should direct its program toward an appropriate compromise between fuel economy and cost using the best available technology to ensure that a market-acceptable production-prototype vehicle can be achieved by 2004 that meets Tier 2 emission standards.

FUEL ECONOMY AND EMISSIONS TRADE-OFFS

Although the emission levels for the new generation of vehicles expected to result from the PNGV program were not quantified in goals 1, 2 or 3, it was stated that they would be the emission standards in place for 2004 for a PNGV-type vehicle. At the outset of the PNGV program, the emission design targets were 0.20 g/mile for NO_x and 0.04 g/mile for PM. In October 1997, almost coincidentally with the PNGV technology selection process (the downselect process), PNGV established new research targets of 0.20 g/mile for NO_x and 0.01 g/mile for PM in anticipation of more stringent regulations. These levels represented an exceedingly difficult challenge and technology stretch for the combustion and after-treatment systems under consideration. Based on the quantitative goal for vehicle fuel economy stated in Goal 3, and because significant advances were being made on the CIDI engine emissions control system, the CIDI engine was selected as the most fuel efficient power train likely to be available in the time frame of the program. A major investigation of the effects of fuel composition on the system efficiencies was also integrated into the program (see Chapter 2). Although the new emission objectives were formidable, the emerging technologies seemed to have the potential to meet them.

On May 13, 1999, EPA announced its proposed Tier 2 emission standards, which introduced another significant tightening of the emission targets for the PNGV vehicle. The proposed standard mandated fleet averages of 0.07 g/mile NO_x and 0.01 g/mile PM, including light trucks and SUVs weighing 8,500 lbs or less. Thus, to meet the Tier 2 targets, significant reductions in NO_x emissions and PM emissions would be required compared to the targets in place at the outset of the program. Although PNGV continues to work toward improving fuel economy, almost all of the CIDI/4SDI program resources have been shifted to investigating advanced emission after-treatment technologies because PNGV believes (and the committee concurs) that no combustion engine will be able to meet the Tier 2 emission levels without them. The Tier 2 standard was finalized and announced by the federal administration on December 21, 1999. Meeting the Tier 2 standards for diesel engines will likely require new catalytic materials and new emissions control concepts.

These Tier 2 emission standards are clearly "technology-forcing" regulations. The data on which EPA claims to have established their feasibility are not statistically significant (Federal Register, 1999). The loss in fuel economy consequent to meeting these standards will depend on the effectiveness and cost of technologies yet to be developed. EPA has acknowledged that reduced sulfur content of the fuel, for both gasoline and diesel engines, will be necessary for the efficiency and durability of new after-treatment systems, but the mechanism for reducing sulfur content and the ultimate level of sulfur have not been determined, although EPA recently proposed a requirement for diesel fuel of no greater than 15 ppm.

Although emissions of greenhouse gases to the atmosphere are not currently regulated, many concerns have been raised about their potential for contributing to climate change. If hydrocarbon fuels are the source of energy for the vehicle power train (a CIDI or any other engine), the most efficient power train will also emit the least amount of carbon dioxide, which would also help reduce the emissions of greenhouse gases to the atmosphere. Concerns about reducing emissions of greenhouse gases has stimulated a large increase in the use of diesel engines in Europe and the rest of the world, for both trucks and passenger cars.

The new standards will certainly require that the PNGV program reassess the relative merits of the CIDI engine and the gasoline spark-ignition engine as its power plant of choice and will certainly reduce the likelihood of meeting the 80 mpg fuel economy goal. PNGV believes that, in the time frame of the program, the best hope of reaching the fuel economy goal of 80 mpg is with a CIDI engine. However, the engine most likely to meet the Tier 2 emissions standard is the gasoline homogeneous stoichiometric combustion engine, for which the exhaust-gas after-treatment system is the most advanced. Unfortunately, this engine also has the lowest energy efficiency of the candidate engines under investigation. The ultimate comparison between the efficiency of these two engines will be determined only when both systems have been developed and certified to meet the Tier 2 standards.

Fuel-cell vehicles have the potential for meeting future emission require-ments, as well as providing gains in energy conversion efficiency, but not by the 2004 milestone of the PNGV program. Many technical issues for fuel-cell vehicles remain to be solved, including reducing the high cost of the technology. The level of emissions from the fuel-cell vehicle depends on whether an onboard fuel reformer is used to produce hydrogen, and which fuel is used in the reformer as the source of hydrogen (see the committee's fourth and fifth reports for more discussion [NRC, 1998, 1999]). No meaningful data are available to assess these systems because most of the expected emissions will come from start-up and transients, and no complete and integrated systems have been operated to date. An onboard reformer would substantially reduce the overall efficiency of the fuel-cell system. Storage of hydrogen fuel would result in a more efficient power plant, but the infrastructure for dispensing hydrogen for large-scale automotive use has not been defined and could not be put in place by 2004.

Recommendation

Recommendation. PNGV should quantify the trade-off between efficiency and emissions for the power plants under consideration. The PNGV systems-analysis team should develop and validate vehicle emissions models of sufficient sophis-tication to provide useful predictions of the emissions potential for a variety of engines (e.g., the compression-ignition direct-injection engine, the gasoline direct-injection engine) and exhaust-gas after-treatment systems in various hybrid elec-tric vehicle configurations. The models could be used to help PNGV evaluate the feasibility of meeting the Environmental Protection Agency's Tier 2 emissions levels and the fuel economy levels that could be achieved with various vehicle system configurations. The impact on greenhouse gas emissions also should be determined. These data should then be used to help establish a plan for the next phase of the program.

FUEL ISSUES

Reducing automotive fuel consumption in the transportation sector will require the widespread availability of affordable vehicles and fuels that meet the requirements of these vehicles. As the committee pointed out in previous reports, the primary vehicle power plants under consideration by the PNGV program could have wide-ranging effects on the fuels industry (NRC, 1998, 1999). Thus, to ensure the availability of the required fuels, the petroleum industry must be involved in the program in a timely fashion.

Each of the vehicle power-plant options under development in the PNGV program for achieving the fuel economy goal of up to 80 mpg has important implications for the composition of fuel. The primary power plant systems under consideration are the CIDI engine in an HEV configuration and fuel cells.

Fuels for CIDI Engines

The CIDI engine has substantial benefits in fuel economy compared to gasoline engines, but emissions of PM and NO_x are a serious problem. Modern cylinder-injection, high rail pressures, and closed-loop controls can dramatically reduce these emissions but most likely cannot meet the EPA Tier 2 standards with current diesel fuel, even with PM traps and NO_x absorbers. Therefore, PNGV is investigating changes in fuel characteristics, such as volatility, aromatic content, and sulfur content, as well as the addition of oxygenates. A reduction of sulfur concentrations in the fuel to less than 10 ppm alone would require major modifications to refineries to produce significant commercial quantities of fuel and would increase the cost of diesel fuel. The low sulfur in the fuel would, in turn, improve the effectiveness of vehicle emission control systems. Thus, there are trade-offs between fuel and vehicle costs.

Refineries are highly interconnected, interrelated systems, and changes in one product output affects other product outputs. In U.S. refineries today, about 50 percent of crude oil is converted to gasoline. This is accomplished by a combination of recovering gasoline fractions found naturally in crude oil, cracking high molecular weight streams to gasoline, converting low molecular weight streams to gasoline, and upgrading the gasoline to meet requirements, such as octane and sulfur level. On a volumetric basis, the amount of diesel fuel produced is about 30 percent of the amount of gasoline produced. If significantly more diesel fuel relative to gasoline were required, processing schemes would have to be modified. In addition, less hydrogen would be produced, and, if the sulfur level in diesel fuel were restricted, more hydrogen would be required to remove the sulfur as hydrogen sulfide. In many refineries, this combination of circumstances would most likely lead to a hydrogen deficiency and the need to build additional hydrogen plants.

In the face of these challenges, the PNGV program has been devoting more attention to fuel composition issues and working with individual petroleum companies but has not yet established an overall coordinated mechanism to determine the commercial trade-offs between engine systems and fuel compositions. For example, a PNGV/oil company *ad hoc* test program has been initiated to identify diesel fuels and fuel properties that could facilitate successful use of the CIDI engine in the United States. Three auto companies, three oil companies, and DOE are involved in this program (see Chapter 2). In addition, individual auto/oil company programs have been initiated, and EPA is pursuing programs related to the regulation of fuel composition to improve air quality. CARB also has a program, the California Fuel Cell Partnership, to develop fuels for fuel cells. This program is also looking into fuel choice and infrastructure issues as well.

DOE has drafted a new Ultra Clean Transportation Fuels Program Plan that proposes to involve both energy and auto companies in developing fuels appropriate for advanced CIDI engines and fuel cells (Chalk, 2000). The overall mission of the program is to pursue, in cooperation with industry, technologies and systems for advanced highway vehicles that improve energy security, environmental quality, and U.S. competitiveness. The fuels in the program will include fuels from renewable energy sources, petroleum, coal, and natural gas, and the objectives include reducing smog-forming emissions and greenhouse gas emissions. As DOE finalizes the plan for this program, the committee encourages DOE to focus on longer term objectives, such as the production of biofuels, and to defer to industry on shorter term objectives, such as the production of gasoline and diesel fuels with specific sulfur levels and engine optimization studies.

Programs by other government agencies could provide useful information about changes in fuel composition. Based on past experience, significant modifications required in the marketplace will take more than five years to define the commercial fuel specifications and design and build the necessary refinery facilities. For example, for gasoline-powered automobiles, the automotive and fuels industries participated in the Auto-Oil Air Quality Improvement Research Program, which investigated the effects of gasoline composition and automotive systems on automotive emissions. The cost was $40 million, and it took more than five years to complete (AQIRP, 1997). Additional time was required to implement commercial changes based on the research program.

Up to now, PNGV's top priority has been the definition of automotive systems that could achieve the goals of the program. Although this was the correct approach in the early stages of the program, it is now critically important that fuel issues be strategically addressed with the involvement of the petroleum industry. Otherwise, because of the lead time required to manufacture modified fuels, commercialization of the technologies being developed could be delayed.

Hydrogen for Fuel Cells

Because most fuel cells under consideration run on hydrogen,[1] the method of providing this fuel is a subject of intense investigation. Two approaches are under study: (1) onboard storage of hydrogen, either as compressed gas (5,000 psi), in a chemical hydride, or as a liquid in a cryogenic tank at very low temperature (e.g., −253°C);[2] and (2) the indirect storage of hydrogen as conventional liquid fuel (gasoline or methanol) from which the hydrogen is "extracted"

---

[1] Fuel cells that run directly on methanol are also being developed.

[2] Liquid hydrogen is an option that has not received much attention because it requires extremely low temperatures. Ogden (1999) has shown that the distribution of liquid hydrogen to service stations and subsequent vaporization would be one of the most expensive ways of delivering gaseous hydrogen to vehicles.

TABLE 5-1 Infrastructure Investment for the Production and Distribution of Hydrogen and Methanol

	Gaseous Hydrogen ($ billions)		Methanol ($ billions)	
	Limited[a]	Wide[b]	Limited[a]	Wide[b]
Production	10	230–400	3.2	84
Distribution	7.7	175	0.36	9

Note: Feedstock for production of hydrogen and methanol is natural gas.

[a] Limited penetration = 70,000 barrels/day gasoline equivalent in 2015.
[b] Wide penetration = 1,600,000 barrels/day gasoline equivalent in 2030.

Source: Wang et al., 1997.

in an onboard fuel reformer and sent to the fuel cell on demand. In the first approach, the hydrogen source must be part of the fuel infrastructure. Gasoline, of course, has the advantage of being widely available in existing service stations. Although future gasoline will probably be a naptha-type hydrocarbon, it could be made available through existing service stations.

Methanol is another fuel under consideration by a number of companies. Methanol reforming technology is more advanced than the technology for reforming gasoline and has the advantage of working at a much lower temperature (260°C) than gasoline reforming (more than 600°C). Although methanol is not widely available, given adequate time to establish a distribution system and additional methanol production facilities, it could be distributed to existing service stations and dispensed to vehicles. However, issues related to its corrosive properties and potential public health effects would have to be investigated and addressed.

Table 5-1 summarizes the results of a study that developed capital cost estimates for the large-scale production and distribution of six fuels, including gaseous hydrogen and methanol (Wang et al., 1997). Based on investment estimates in Table 5-1, Kalhammer et al. (1998) estimated that infrastructure investment per vehicle would be between $3,500 and $6,700 for hydrogen and between $710 and $820 for methanol.[3]

These high costs provide a large incentive for the development of an onboard gasoline reformer, and PNGV has devoted significant resources to this objective.

---

[3] Infrastructure costs based on projected vehicle penetration were estimated for daily delivery of hydrogen with the energy equivalence of 70,000 barrels of gasoline in 2015 and of 1.6 million barrels of gasoline in 2030. The hydrogen-fueled vehicles were assumed to be driven 14,000 miles per year at a fuel economy equivalent to either 60 or 80 mpg (gasoline). The infrastructure costs given in the text cover the range of costs based on these assumptions.

However, problems abound, and no cost effective, practical system is in sight at this point. First, an onboard gasoline reformer would be expensive and would require an excessive amount of platinum. Second, start-up time for the reformer system would be on the order of minutes rather than seconds. Third, an integrated reformer/fuel-cell system has not yet been tested to evaluate operating dynamics. Fourth, gasoline molecules form soot in the reformer, suggesting that the hydrocarbon stream would have to be lighter than conventional gasoline and that sulfur levels would have to be very low.

In light of the infrastructure costs of hydrogen storage and the status of the onboard reforming program, the committee feels that PNGV should assess approaches for generating hydrogen at service stations by, for example, reforming natural gas or gasoline. In effect, this would replace the mobile reformer on each car with larger, stationary reformers that could provide fuel for many vehicles. The larger size of these units would provide additional design flexibility to deal with important technical issues, such as platinum requirements and the operability of the system with gasoline. In addition, with this option, hydrogen could be generated from natural gas, a preferred reformer feedstock that would minimize cost. Finally, this approach would avoid the infrastructure costs of a gaseous hydrogen distribution system and, therefore, could be implemented more quickly.

When investigating this option, PNGV should take advantage of the expertise of the National Aeronautics and Space Administration, which is developing new technology for hydrogen storage, leak detection, and firefighting. PNGV should also take advantage of the expertise of the petroleum and chemical industries, which have extensive experience in generating, storing, transporting, and using hydrogen in refineries and chemical plants.

Recommendations

Recommendation. Defining automotive system/fuels trade-offs and establishing the basis for planning for supplying required fuels as higher efficiency vehicles become commercially available will require extensive cooperation among automotive and petroleum industry representatives at all levels of responsibility. PNGV should expand and strengthen its cooperative efforts with the petroleum industry, including issues related to fuels for fuel cells. Government leadership will be necessary to initiate this cooperative effort and provide incentives for petroleum company involvement.

Recommendation. PNGV should undertake a study to assess the opportunities and costs for generating hydrogen for fuel cells at existing service stations and storing it on board vehicles and compare the feasibility, efficiency, and safety of this option with onboard fuel reforming. This study will help PNGV determine how much additional effort should be devoted to the development of onboard fuel reforming technologies.

6

PNGV's Response to the Fifth Report

In its previous five reviews, the National Research Council Standing Committee to Review the Research Program of the PNGV made a number of recommendations, which are documented in published reports (NRC, 1994, 1996, 1997, 1998, 1999). In the fifth report, the committee made specific recommendations related to each of the technologies under development and general recommendations for the program as a whole. Appendix B contains a letter from PNGV to the committee chairman documenting PNGV's responses to the major recommendations in the Executive Summary of the fifth report (NRC, 1999). PNGV's responses indicate that the PNGV agrees, for the most part, with the committee's recommendations and has responded favorably to the committee's suggestions. Discussions of PNGV's responses to the technical suggestions and recommendations in the fifth report are incorporated in the corresponding technical sections in chapters 2 and 3.

The following comments relate to two of the major recommendations from the fifth report (the responses below can also be found in Appendix B):

Recommendation. The federal government agencies involved in the PNGV program should review how future emissions requirements (especially NO_x and particulates), fuel economy, and carbon dioxide emissions, as well as fuel quality, will affect the choice of the compression-ignition direct-injection engine as the most promising short-term combustion engine technology; a program plan that responds to that assessment should be developed. The PNGV, especially the U.S. Department of Energy and the Environmental Protection Agency, should work closely with the California Air Resources Board on these issues.

PNGV's Response. In response to pending Tier 2 federal emissions regulations, the PNGV partners have adopted a more aggressive R&D program to reduce NO_x and particulate emissions from the CIDI engine while maintaining its inherently high efficiency and low carbon emissions. Within the Low Emissions Partnership (LEP) lean NO_x catalyst and non-thermal plasma cooperative research efforts, a new goal of 90 percent or greater NO_x conversion is being considered and selective catalytic reduction (SCR) is being added to the NO_x reduction strategies. Fuels and aftertreatment programs continue to be more closely linked to minimize in-cylinder emissions while optimizing the performance of new clean fuels as emission control reductants. Two new three-year cooperative agreements to develop emission control systems for the Ford and DaimlerChrysler PNGV engines have brought the expertise of catalyst suppliers more directly into the program. In addition, the DOE has been working with EPA, Engine Manufacturers Association (EMA) members and emission control manufacturers through the Diesel Emission Control Sulfur Effects (DECSE) project to determine the effects of sulfur in diesel on emission control devices.

The committee is not convinced that PNGV has conducted an adequate analysis of how emission requirements, fuel economy, carbon dioxide emissions, and fuel quality will affect the choice of the CIDI engine. As the committee noted in previous chapters, meeting the Tier 2 emission standards with a CIDI engine will be a formidable challenge, especially in the time frame of the PNGV program. The committee understands that the setting of the Tier 2 standards, although not supported by robust statistical analysis and analytical data, was driven by broad considerations of public health and air quality (Federal Register, 1999). Given the impact of these standards, the committee's recommendation from the fifth report has become even more important. However, PNGV did not provide a program plan that responds to the recommended assessment.

Recommendation. Without compromising proprietary information of the USCAR partners, the PNGV should conduct in-depth cost analyses and use the results to guide subsystem and vehicle affordability studies.

PNGV's Response. PNGV agrees that it needs to conduct in-depth cost analyses. This necessarily occurs at the vehicle and subsystem levels. The vehicle-level analyses, completed in 1998, are company and configuration specific and thus can only be reviewed in the individual company proprietary sessions. The PNGV directors used the information from their proprietary analyses to jointly develop subsystem-level cost targets for generic fuel cell and hybrid-electric vehicles.

Each technical team used the targets from these generic models to optimize their respective subsystems. An update of the subsystem cost analyses will be presented by the Technical Teams at the November 1999 Collaborative Peer Review.

The projected production costs of all candidate PNGV vehicles presented to the committee far exceed the cost objectives of the program, which are for a vehicle with equivalent costs of ownership, adjusted for economics, to a 1994 family sedan. The committee received cost estimates for some candidate vehicles from some of the PNGV participants. Because cost is a critical factor, the committee has requested a detailed review of future cost projections as soon as possible. These projections are extremely important as the program moves toward the development of the production prototype vehicle and as cost reduction becomes an increasing priority.

The committee is also concerned that PNGV has not provided a systematic analysis of the relationship between proprietary cost analyses and the cost targets for vehicle technology subsystems. If the subsystem cost targets are credible, then they should add up to an affordable cost for the vehicle.

References

AQIRP (Auto/Oil Air Quality Improvement Research Program). 1997. AQIRP Program Final Report. Detroit, Mich.: Auto/Oil Air Quality Improvement Research Program.

Askari, E. 2000. At Last the Supercars Arrive. Detroit Free Press, Section 1C, January 6, 2000.

Brown, S.F. 2000. Tough Composites Hit the Road. Fortune Technology Magazine 141(2): 110B. Available on line at: *http:library.northernlights.com*

Chalk, S. 2000. Overview of DOE Fuels Activities. Presentation to the Standing Committee to Review the Research Program of the Partnership for a New Generation of Vehicles, National Academy of Sciences, Washington, D.C., February 24, 2000.

DaimlerChrysler. 2000. ESX3, Converging on Reality. Press Kit. Available from DaimlerChrysler Communications, 800 Chrysler Drive, Auburn Hills, MI 48326-2766. Also available on line at: *www.media.daimlerchrysler.com*

DOE (U.S. Department of Energy). 1997. Scenarios of U.S. Carbon Reductions: Potential Impacts of Energy-Efficient and Low-Carbon Technologies to 2010 and Beyond. Washington, D.C.: U.S. Department of Energy, Interlaboratory Working Group on Energy-Efficient and Low-Carbon Technologies.

EPA (Environmental Protection Agency). 2000. Proposed Heavy-Duty Engine and Vehicle Standards and Highway Diesel Fuel Sulfur Control Requirements. Regulatory Announcement EPA-F-00-022 (May). Washington, D.C.: Office of Transportation and Air Quality, Environmental Protection Agency.

Evans, L. 2000. Causal Influence of Mass and Size on Driver Fatality Risk. Report R&D-9035. Warren, Mich.: General Motors Research and Development Center.

Federal Register. 1999. Control of Air Pollution from New Motor Vehicles: Proposed Tier 2 Motor Vehicle Emissions Standards and Gasoline Sulfur Control Requirements: Regulatory Impact Analysis (May 13, 1999). Environmental Protection Agency Notice of Proposed Rule Making. 64: 92.

Ford. 2000. Technology Press Release. Distributed at the North American International Automobile Show, Detroit, Michigan, January 11, 2000. Available from Jon Harmon or Brendan Prebo, Ford Motor Company Technology Public Affairs, The American Road, Dearborn, MI 48126.

Frei, P., R. Kaeser, R. Hafner, M. Schmidt, A. Dragan, M. Wingeier, H. Muser, P.F. Niederer, and F.H. Walz. 1997. Crashworthiness and Compatibility of Low Mass Vehicles in Collisions. Society of Automotive Engineers (SAE) Paper 970122. Warrendale, Pa.: Society of Automotive Engineers.

GM (General Motors Corporation). 2000. gm://driving.the.next.generation. Press kit distributed at the North American International Automobile Show, Detroit, Michigan, January 11, 2000. Available from Jeff Kuhlman or Jennifer Schmit, GM Communications, or *jeffrey.kuhlman@gm.com* or *jennifer.1.schmit@gm.com*

Haskins, H. 1999. Battery Technical Team Review. Presentation to the Standing Committee to Review the Research Program of the Partnership for a New Generation of Vehicles, USCAR Headquarters, Southfield, Michigan, November 18, 1999.

Jeannes, D., and M. van Schaik. 2000. Steel Body Approaches to Reducing Vehicle Weight. Presentation to the Standing Committee to Review the Research Program of the Partnership for a New Generation of Vehicles, National Academy of Sciences, Washington, D.C., February 23, 2000.

Jewett, D. 1997. Ford unveils light car of the future. Automotive News 5738: 46.

Kalhammer, F., P.R. Prokopius, V.P. Roan, and G.E. Voecks. 1998. Status and Prospects of Fuel Cells as Automobile Engines. Sacramento, Calif.: State of California Air Resources Board.

Low, S., T. Asmus, R. Peterson, K. Howden, and K. Hellman. 1999. 4SDI Technical Team Review. Presentation to the Standing Committee to Review the Research Program of the Partnership for a New Generation of Vehicles, USCAR Headquarters, Southfield, Michigan, November 18, 1999.

Malcolm, R., J. Merritt, D. Hamilton, B. Murty, and A. Gibson. 1999. Electrical and Electronics Technical Team Review. Presentation to the Standing Committee to Review the Research Program of the Partnership for a New Generation of Vehicles, USCAR Headquarters, Southfield, Michigan, October 27, 1998.

Milliken, J. 1999. Fuel Cell Technical Team Review. Presentation to the Standing Committee to Review the Research Program of the Partnership for a New Generation of Vehicles, USCAR Headquarters, Southfield, Michigan, November 17, 1999.

Niederer, P.F. 1993. Occupant Safety of Low Mass Vehicles. Society of Automotive Engineers (SAE) Paper 933107. Warrendale, Pa.: Society of Automotive Engineers.

National Research Council (NRC). 1992. Automotive Fuel Economy. Energy Engineering Board. Washington, D.C.: National Academy Press.

NRC. 1994. Review of the Research Program of the Partnership for a New Generation of Vehicles. Board on Energy and Environmental Systems and Transportation Research Board. Washington, D.C.: National Academy Press.

NRC. 1996. Review of the Research Program of the Partnership for a New Generation of Vehicles, Second Report. Board on Energy and Environmental Systems and Transportation Research Board. Washington, D.C.: National Academy Press.

NRC. 1997. Review of the Research Program of the Partnership for a New Generation of Vehicles, Third Report. Board on Energy and Environmental Systems and Transportation Research Board. Washington, D.C.: National Academy Press.

NRC. 1998. Review of the Research Program of the Partnership for a New Generation of Vehicles, Fourth Report. Board on Energy and Environmental Systems and Transportation Research Board. Washington, D.C.: National Academy Press.

NRC. 1999. Review of the Research Program of the Partnership for a New Generation of Vehicles, Fifth Report. Board on Energy and Environmental Systems and Transportation Research Board. Washington, D.C.: National Academy Press.

Ogden, J. 1999. Prospects for Building a Hydrogen Energy Infrastructure. Annual Review of Energy and the Environment 24: 227–268.

Office of Technology Assessment (OTA). 1995. Advanced Automotive Technology: Visions of a Super-Efficient Family Car. OTA-ETI-638. Washington, D.C.: Government Printing Office.

PCAST (President's Committee of Advisors on Science and Technology). 1997. Federal Energy Research and Development for the Challenges of the Twenty-First Century (November 5). Washington, D.C.: Executive Office of the President.

PNGV (Partnership for a New Generation of Vehicles). 1995. Partnership for a New Generation of Vehicles Program Plan (draft). Washington, D.C.: U.S. Department of Commerce, PNGV Secretariat.

PNGV. 1999a. Technical Roadmap (updated draft, September, 1999). Southfield, Mich.: PNGV/USCAR.

PNGV. 1999b. Answers from the PNGV to questions from the Standing Committee to Review the Partnership for a New Generation of Vehicles, December 17, 1999.

Pryweller, J. 1999a. D/C's plastic SUV commands attention. Automotive News 5802: 26B.

Pryweller, J. 1999b. Decoma joins D/C suppliers developing plastic car. Automotive News 5800: 20.

Robinson, A. 2000. DCX's high-mileage concept car gets 72 mpg. Automotive News 5863: 3.

Ryan, T.W., J. Buckingham, L.G. Dodge, and C. Olikara. 1998. The Effect of Fuel Properties on Emissions from a 2.5 gm NO_x Heavy-Duty Diesel Engine. Society of Automotive Engineers (SAE) Paper 9982491. Warrendale, Pa.: Society of Automotive Engineers.

Schultz, R.A. 1999. Aluminum for Lightweight Vehicles: An Objective Look at the Next 10 Years to 20 Years. Presented at the Metal Bulletin 14th International Aluminum Conference, Montreal, Canada, September 15, 1999. Available from Richard A. Schultz, Ducker Research Co., Inc., or *richards@drucker.com*

Sherman, A. 1998. Materials Team Review. Presentation to the Standing Committee to Review the Research Program of the Partnership for a New Generation of Vehicles, USCAR Headquarters, Southfield, Michigan, October 26, 1998.

Simunovic, S., and J. Carpenter. 1999. Impact Modeling of Lightweight Automotive Structures. Oak Ridge, Tenn.: Oak Ridge National Laboratory. Available on line at: *http://www-explorer.ornl.gov*

Sissine, F. 1996. The Partnership for a New Generation of Vehicles (PNGV). Report No. 96-191 SPR. Washington, D.C.: Congressional Research Service.

Stuef, B. 1997. Vehicle Engineering Team Review. Presentation to the Standing Committee to Review the Research Program of the Partnership for a New Generation of Vehicles, USCAR Headquarters, Southfield, Michigan, October 14, 1997.

Takatori, Y., Y. Mandokoro, K. Akihama, K. Nakakita, Y. Tsukadaki, S. Iguchi, L. Yeh, and T. Dean. 1998. Effect of Hydrocarbon Molecular Structure on Diesel Exhaust Emissions. Part 2: Effect of Branched and Ring Structures of Paraffins on Benzene and Soot Formation. Society of Automotive Engineers (SAE) Paper 982495. Warrendale, Pa.: Society of Automotive Engineers.

Tanaka, S., and H. Takizawa. 1998. Effect of Fuel Compositions on PAH in Particulate Matter from DI Diesel Engine. Society of Automotive Engineers (SAE) Paper 982648. Warrendale, Pa.: Society of Automotive Engineers.

Wall, J.C., and S.K. Hoekman. 1984. Fuel Composition Effects on Heavy-Duty Diesel Particulate Emissions. Society of Automotive Engineers (SAE) Paper 841364. Warrendale, Pa.: Society of Automotive Engineers.

The White House. 1993. Historic Partnership Forged with Automakers Aims for Threefold Increase in Fuel Efficiency in as Soon as Ten Years. Washington, D.C.: The White House.

ULSAB (UltraLight Steel Auto Body). 1999. Available on line at: *http://www.ulsab.org*

Wang, M., K. Stork, A. Vyas, M. Mintz, M. Singh, and L. Johnson. 1997. Assessment of PNGV Fuels Infrastructure Phase 1 Report: Additional Capital Needs and Fuel-Cycle Energy and Emissions Impacts. Report No. ANL/ESD/TM-140. Argonne, Ill.: Argonne National Laboratory.

Appendices

APPENDIX A

Biographical Sketches of Committee Members

Trevor O. Jones, chair, is chairman and chief executive officer (CEO) of Biomec, Incorporated, a biomedical device company. He was formerly chairman of the board of Echlin, Incorporated, a supplier of automotive components primarily to the after-market. Mr. Jones is also chairman and CEO of International Development Corporation, a private management consulting company that advises automotive supplier companies on strategy and technology. He was chair, president, and CEO (retired) of Libbey-Owens-Ford Company, a major manufacturer of glass for automotive and construction applications. Previously, he was an officer of TRW, Incorporated, serving as vice president of engineering in the company's Automotive Worldwide Sector and group vice president, Transportation Electronics Group. Prior to joining TRW, he was employed by General Motors (GM) in many aerospace and automotive executive positions, including director of GM Proving Grounds; director of the Delco Electronics Division, Automotive Electronic and Safety Systems; and director of GM Advanced Product Engineering Group. Mr. Jones is a life fellow of the American Institute of Electrical and Electronics Engineers and has been cited for "leadership in the application of electronics to the automobile." He is also a fellow of the American Society of Automotive Engineers, a fellow of the British Institution of Electrical Engineers, a fellow of the Engineering Society of Detroit, a registered professional engineer in Wisconsin, and a chartered engineer in the United Kingdom. He holds many patents and has lectured and written on automotive safety and electronics. He is a member of the National Academy of Engineering (NAE) and a former commissioner of the National Research Council (NRC) Commission on Engineering and Technical Systems. Mr. Jones has served on several other NRC study committees, including the Committee for a Strategic Transportation Research Study on

97

Highway Safety, and chaired the NAE Steering Committee on the Impact of Products Liability Law on Innovation. He holds an HNC (Higher National Certificate) in electrical engineering from Aston Technical College and an ONC (Ordinary National Certificate) in mechanical engineering from Liverpool Technical College.

Craig Marks, vice chair, is president of Creative Management Solutions. He has been adjunct professor in both the College of Engineering and the School of Business Administration at the University of Michigan and co-director of the Joel D. Tauber Manufacturing Institute. Dr. Marks was also president of the Environmental Research Institute of Michigan. He is a retired vice president of technology and productivity for AlliedSignal Automotive, where he was responsible for product development; manufacturing; quality; health, safety, and environment; communications; and business planning. Previously, in TRW's Automotive Worldwide Sector, Dr. Marks was vice president for engineering and technology and later vice president of technology at TRW Safety Restraint Systems. Prior to joining TRW, he held various positions at GM Corporation, including executive director of the engineering staff; assistant director of advanced product engineering; engineer in charge of power development; electric-vehicle program manager; supervisor for long-range engine development; and executive director of the environmental activities staff. He is a member of the NAE and a fellow of the Society of Automotive Engineers. Dr. Marks received his B.S.M.E., M.S.M.E., and Ph.D. in mechanical engineering from the California Institute of Technology.

William Agnew retired as director, Programs and Plans, General Motors Research Laboratories in 1989. He served in the Manhattan District from 1944 to 1946, and attended Purdue University from 1946 to 1952. From 1952 to 1989, he held a number of positions at GM Research Laboratories, including department head, Fuels and Lubricants; head, Emissions Research Department; technical director, Engine Research, Engineering Mechanics, Mechanical Research, Fluid Dynamics, and Fuels and Lubricants Departments; technical director, Biomedical Science, Environmental Science, Societal Analysis, and Transportation Research Departments. A member of the NAE, Dr. Agnew's technical expertise spans internal combustion engines, gas turbines, engine performance, automotive air pollution, and automotive power plants. He has a Ph.D. in mechanical engineering from Purdue University.

Alexis T. Bell is professor of chemical engineering, University of California, Berkeley. He has also held the positions of dean, College of Chemistry, and chairman, Department of Chemical Engineering. The emphasis of his research is on heterogeneous catalysis and the relationship between catalyst composition and structure and catalyst performance on the molecular level. He is a recipient of the Curtis W. McGraw Award for Research, American Association of Engineering

Education; the Professional Progress Award, American Institute of Chemical Engineers; the Paul H. Emmett Award in Fundamental Catalysis, the Catalysis Society; and the R.H. Wilhelm Award in Chemical Reaction Engineering, American Institute of Chemical Engineers. He is a member of the NAE and a fellow of the American Association for the Advancement of Science. He received his Sc.D. in chemical engineering from the Massachusetts Institute of Technology (MIT).

W. Robert Epperly is president of Epperly Associates, Incorporated, a consulting firm. From 1994 to 1997, he was president of Catalytica Advanced Technologies, Incorporated, a company that develops new catalytic technologies for the petroleum and chemical industries. Prior to joining Catalytica, he was general manager of Exxon Corporate Research and director of the Exxon Fuels Research Laboratory. After leaving Exxon, he was CEO of Fuel Tech N.V., a company that develops new combustion and air pollution control technologies. Mr. Epperly has written or co-authored more than 50 publications on technical and managerial topics, including two books, and has 38 U.S. patents. He has extensive experience in fuels, fuel cells, engines, catalysis, air pollution control, and the management of research and development programs. He received an M.S. degree in chemical engineering from Virginia Polytechnic Institute.

David E. Foster is professor of mechanical engineering, University of Wisconsin, Madison, and former director of the Engine Research Center, which has won two center of excellence competitions for engine research and has extensive facilities for research on internal combustion engines. A member of the faculty at the University of Wisconsin since he completed his Ph.D., Dr. Foster teaches and conducts research in thermodynamics, fluid mechanics, internal combustion engines, and emission formation processes. His work has focused specifically on perfecting the application of optical diagnostics in engine systems and the incorporation of simplified or phenomenological models of emission formation processes into engineering simulations. He has published more than 60 technical articles in this field throughout the world and for leading societies in this country. He is a recipient of the Ralph R. Teetor Award, the Forest R. McFarland Award, and the Lloyd L. Withrow Distinguished Speaker Award of the Society of Automotive Engineers. He is a registered professional engineer in the State of Wisconsin and has won departmental, engineering society, and university awards for his classroom teaching. He received a B.S. and M.S. in mechanical engineering from the University of Wisconsin and a Ph.D. in mechanical engineering from the MIT.

Norman A. Gjostein is currently clinical professor of engineering, University of Michigan-Dearborn, where he teaches courses in materials engineering. He retired from Ford Research Laboratory as director, Manufacturing and Materials Research Laboratory, which includes research in advanced materials, manufacturing systems, and computer-aided engineering. He has directed a variety of advanced research programs, including the development of lightweight metals, composite materials, sodium-sulfur batteries, fiber-optic multiplex systems, and smart sensors. He has pioneered studies in surface science and discovered a number of new surface structures that are still under investigation. He is a member of the NAE and NRC Commission on Engineering and Technical Systems, a fellow of the Engineering Society of Detroit (ESD) and the American Society of Metals (ASM), and a recipient of the ASM's Shoemaker Award and ESD's Gold Award. Dr. Gjostein has a B.S. and M.S. in metallurgical engineering from the Illinois Institute of Technology and a Ph.D. in metallurgical engineering from Carnegie-Mellon University.

David F. Hagen spent 35 years with Ford Motor Company, where his position prior to retirement was general manager, alpha simultaneous engineering, Ford Technical Affairs. Under his leadership, Ford's alpha activity, which involves the identification, assessment, and implementation of new product and process technologies, evolved into the company's global resource for leading-edge automotive products, processes, and analytic technologies. Mr. Hagen led the introduction of the first domestic industry feedback electronics, central fuel metering, full electronic engine controls, and numerous four-cylinder, V6, and V8 engines. Based on his work on Ford's modern engine families, he was awarded the Society of Automotive Engineers E.N. Cole Award for Automotive Engineering Innovation in 1998. Mr. Hagen received his B.S. and M.S. in mechanical engineering from the University of Michigan. He is a fellow of the Engineering Society of Detroit and a member of the Society of Automotive Engineers. He is currently serving on the engineering advisory boards of both Western Michigan University and the University of Michigan-Dearborn, as well as the board of the Rackham Engineering Foundation.

John B. Heywood is Sun Jae Professor of Mechanical Engineering at the Massachusetts Institute of Technology and director of the Sloan Automotive Laboratory. Dr. Heywood's research interests have focused on understanding and explaining the processes that govern the operation and design of internal combustion engines and their fuel requirements. His major areas of research include engine combustion, pollutant formation, operating and emissions characteristics, and fuel requirements of automotive and aircraft engines. He has been a consultant to Ford Motor Company, Mobil Research and Development Corporation, and several other industry and government organizations. He received the U.S. Department of Transportation 1996 Award for the Advancement of Motor Vehicle Research and Development, as well as several awards from the Society of Automotive Engineers (SAE) and other organizations. Dr. Heywood has a Ph.D. in mechanical engineering from the Massachusetts Institute of Technology and an Sc.D. from Cambridge University. He is a fellow of SAE and a member of the NAE.

Fritz Kalhammer is a consultant for the Electric Power Research Institute's (EPRI's) Strategic Science and Technology and Transportation Groups. He was co-chair of the California Air Resources Board's Battery Technical Advisory Panels on electric vehicle batteries, and he recently chaired a similar panel to assess the prospects of fuel cells for electric vehicle propulsion. He has been vice president of EPRI's Strategic Research and Development and established the institute's programs for energy storage, fuel cells, electric vehicles, and energy conservation. Before joining EPRI, he directed electrochemical energy conversion, storage, and process research and development at Stanford Research Institute (now SRI International), conducted research in solid-state physics at Philco Corporation, and conducted research in catalysis at Hoechst, in Germany. He has a Ph.D. in physical chemistry from the University of Munich.

John G. Kassakian is professor of electrical engineering and director of the MIT Laboratory for Electromagnetic and Electronic Systems. His expertise is in the use of electronics for the control and conversion of electrical energy, industrial and utility applications of power electronics, electronic manufacturing technologies, and automotive electrical and electronic systems. Prior to joining the MIT faculty, he served in the U.S. Navy. Dr. Kassakian is on the boards of directors of a number of companies and has held numerous positions with the Institute of Electrical and Electronics Engineers (IEEE), including founding president of the IEEE Power Electronics Society. He is a member of the NAE, a fellow of the IEEE, and a recipient of the IEEE's William E. Newell Award for Outstanding Achievements in Power Electronics (1987), the IEEE Centennial Medal (1984), and the IEEE Power Electronics Society's Distinguished Service Award (1998). He has an Sc.D. in electrical engineering from MIT.

Harold H. Kung is professor of chemical engineering at Northwestern University and was director of the university's Center for Catalysis and Surface Science. His areas of research include surface chemistry, catalysis, and chemical reaction engineering. His professional experience includes work as a research chemist at E.I. du Pont de Nemours & Co., Incorporated. He is a recipient of the P.H. Emmett Award and the Robert Burwell Lectureship Award from the North American Catalysis Society, the Herman Pines Award of the Chicago Catalysis Club, the Japanese Society for the Promotion of Science Fellowship, the John McClanahan Henske Distinguished Lectureship of Yale University, and the Olaf A. Hougen Professorship at the University of Wisconsin, Madison. He has a Ph.D. in chemistry from Northwestern University.

John Scott Newman is professor of chemical engineering at the University of California, Berkeley. His research experience is in the design and analysis of electrochemical systems, transport properties of concentrated electrolytic solutions, and various fuel cells and batteries. He has received the Young Author's

Prize from the Electrochemical Society, the David C. Grahame Award, the Henry B. Linford Award, and the Olin Palladium Medal. He is a member of the NAE and a fellow of the Electrochemical Society. He is author of *Electrochemical Systems* (Prentice Hall, 1991), which has been translated into Japanese and Russian, and has been an associate editor of the *Journal of the Electrochemical Society* since 1990. He has a Ph.D. in chemical engineering from the University of California, Berkeley.

Roberta Nichols is retired from Ford Motor Company, where, from 1979 to 1995, she held several positions including: manager, Electric Vehicle External Strategy and Planning Department, North American Automotive Operations; Manager, EV External Affairs, EV Planning and Program Office; manager, Alternative Fuels Department, Environment and Safety Engineering Staff; and principal research engineer, Alternative Fuels Department, Scientific Research Laboratory. She was also a member of the technical staff of the Aerospace Corporation from 1960 to 1979 and has held other industry positions. She is a fellow of the Society of Automotive Engineers, a recipient of the National Achievement Award of the Society of Women Engineers, a recipient of the Clean Air Award for Advancing Air Pollution Technology of the South Coast Air Quality Management District, and a member of the NAE . Her expertise includes alternative fuel vehicles, electric vehicles, internal combustion engines, and strategic planning. She has a Ph.D. in engineering and an M.S. in environmental engineering, University of Southern California, and a B.S. in physics, University of California, Los Angeles.

Vernon P. Roan is director of the Center for Advanced Studies in Engineering and professor of mechanical engineering at the University of Florida, where he has been a faculty member for more than 30 years. Since 1994, he has also been the director of the University of Florida Fuel Cell Research and Training Laboratory. Previously, he was a senior design engineer with Pratt and Whitney Aircraft. Dr. Roan, who has more than 25 years of research and development experience, is currently developing improved modeling and simulation systems for a fuel-cell bus program and working as a consultant to Pratt and Whitney on advanced gas-turbine propulsion systems. His research at the University of Florida has involved both spark-ignition and diesel engines operating with many alternative fuels and advanced concepts. With groups of engineering students, he designed and built a 20-passenger diesel-electric bus for the Florida Department of Transportation and a hybrid-electric urban car using an internal-combustion engine and lead-acid batteries. He has been a consultant to the Jet Propulsion Laboratory monitoring their electric and hybrid vehicle programs. He has organized and chaired two national meetings on advanced vehicle technologies and a national seminar on the development of fuel-cell-powered automobiles and has published numerous technical papers on innovative propulsion systems. He was one of the four members of the Fuel Cell Technical Advisory Panel of the California Air Resources Board, which issued a report in May 1998 regarding the status and outlook for fuel cells for transportation applications. Dr. Roan received his B.S. in aeronautical engineering and his M.S. in engineering from the University of Florida and his Ph.D. in engineering from the University of Illinois.

APPENDIX B

Letter from PNGV

(NOVEMBER 8, 1999)

November 8, 1999

Mr. Trevor Jones
Chair, Standing Committee to Review the
Research Program of PNGV
National Research Council
2101 Constitution Avenue, N.W.
Room # HA 270
Washington, DC 20418

Dear Trevor,

   We want to thank you and the other Standing Committee members for your insightful
and productive review of the Partnership for a New Generation of Vehicles.

   Attached are our comments on the major recommendations from the 5th report. We
agreed with many of your recommendations and we will be ready to discuss these with you at
the upcoming 6th Review.

   You also made a number of recommendations concerning the individual technology
areas. The PNGV Technical Teams will address each of your recommendations during their
discussions with you at the upcoming 6th Review.

   Again we appreciate receiving your valuable analysis as we progress through the
challenges of developing the PNGV technologies and advancing toward our goals.

Sincerely,

Vince Fazio
PNGV Director
Ford

Ron York
PNGV Director
General Motors

Attachment

John Sargent
Director
PNGV Secretariat

Steve Zimmer
PNGV Director
Daimler-Chrysler

Response to the NRC's 5th Peer Review Report Recommendations

RECOMMENDATION: *The PNGV four-stroke direct-injection technical team should develop projections of the performance of compression-ignition direct injection and gasoline direct injection power-train systems, especially comparisons of the estimated emissions and fuel economy for each system. These projections would be a first step toward the quantification of trade-offs between emissions and fuel economy based on current and emerging state-of-the-art technologies.*

The 4SDI Technical Team agrees with the Peer Review Committee's recommendation and continues to quantitatively compare the projected performance (fuel consumption, emissions, cost, weight, performance, etc.) of various 4SDI powertrain alternatives with respect to the PNGV objectives. Some of these comparisons were made on a company proprietary basis, rather than collaboratively. The 4SDI Technical Team has been collaborating with the System Analysis Team on developing projections of the performance of compression-ignition direct-injection and gasoline direct injection power-train systems. Once the systems model has been validated, it would be an appropriate tool to use in quantifying the necessary trade-offs. The teams will present their results at the November 1999 Collaborative Peer Review.

PNGV recognizes there are potential trade-offs between emission and fuel economy targets. We believe EPA's proposed Tier 2 standards presents increased technical challenges to achieving PNGV's fuel economy goal with four-stroke direct-injection technologies. We believe the PNGV program should give increased emphasis to demonstrating 4SDI engines that achieve both Tier 2 emissions levels and the PNGV fuel efficiency goal.

RECOMMENDATION: *The federal government agencies involved in the PNGV program should review how future emissions requirements (especially NOx and particulates), fuel economy, carbon dioxide emissions, as well as fuel quality, will affect the choice of the compression-ignition direct-injection engine as the most promising short-term combustion engine technology; a program plan that responds to that assessment should be developed. The PNGV, especially the U.S. Department of Energy and the Environmental Protection Agency, should work closely with the California Air Resources Board on these issues.*

In response to pending Tier 2 federal emissions regulations, the PNGV partners have adopted a more aggressive R&D program to reduce NOx and particulate emissions from the CIDI engine while maintaining its inherently high efficiency and low carbon emissions. Within the Low Emissions Partnership (LEP) lean NOx catalyst and non-thermal plasma cooperative research efforts, a new goal of 90 percent or greater NOx conversion is being considered and selective catalytic reduction (SCR) is being added to the NOx reduction strategies. Fuels and aftertreatment programs continue to be more closely linked to minimize in-cylinder emissions while optimizing the performance of new clean fuels as emission control reductants. Two new three-year cooperative agreements to develop emission control systems for the Ford and DaimlerChrysler PNGV engines have brought the expertise of catalyst suppliers more directly into the program. In addition, the DOE has been working with EPA, Engine Manufacturers Association (EMA) members and emission control manufacturers through the Diesel Emission Control Sulfur Effects (DECSE) project to determine the effects of sulfur in diesel on emission control devices.

RECOMMENDATION: *A comprehensive mechanism should be established to help define feasible, timely, and compatible fuel and power-plant modifications to meet the PNGV goals. This mechanism will require extensive cooperation among not only automotive and fuels industry participants at all levels of responsibility, but also among technical and policy experts of the relevant government organizations.*

We agree with this recommendation and efforts at various levels of responsibility in the industries have been initiated. At the Deputy Secretary level, the Departments of Energy and Commerce have met with representatives from energy companies and the American Petroleum Institute (API) to discuss a mechanism that will help define compatible fuel, power-plant and emission control combinations to meet PNGV goals. Discussions are ongoing and anticipated input from the API will allow high level cooperation between automotive and fuels industries to move forward; these efforts, however, should be accelerated. At the technical level, the DOE has drafted a Multi-year Program Plan for Advanced Petroleum Based Fuels for CIDI engines. This plan has been discussed in detail with the energy, automotive, emission control and heavy vehicle industries. In addition, the Ad Hoc Auto/Energy Working Group continues to be an excellent forum to bring the automotive and energy industries together to work on technical issues. Testing of advanced fuels in advanced PNGV-type engines is continuing and new projects, which include the evaluation of additional oxygenates, lube oil contribution to PM emissions, and PM toxicity analysis, have been initiated through the Ad Hoc Auto/Energy Working Group.

RECOMMENDATION: *The PNGV should conduct life-cost and performance-cost trade-off studies, as well as materials and manufacturing cost analyses, to determine which battery technology offers the best prospects and most attractive compromises for meeting capital and life-cycle cost targets.*

PNGV agrees with the peer review's assessment. Life-cost and performance-cost trade-off studies are already included in the developers' existing statements of work. The Electrochemical Energy Storage Technical Team will collaborate with the Systems Analysis Tech Team to analyze test data and will develop performance and business models based on the combined costs of the battery and associated power electronics. The Systems Analysis Tech Team will use these models to determine the most attractive battery technology, taking into account performance, life, and cost compromises.

RECOMMENDATION: *Without compromising proprietary information of the USCAR partners, the PNGV should conduct in-depth cost analyses and use the results to guide subsystem and vehicle affordability studies.*

PNGV agrees that it needs to conduct in-depth cost analyses. This necessarily occurs at the vehicle and subsystem levels. The vehicle-level analyses, completed in 1998, are company and configuration specific and thus can only be reviewed in the individual company proprietary sessions. The PNGV directors used the information from their proprietary analyses to jointly develop subsystem-level cost targets for generic fuel cell and hybrid-electric vehicles.

Each technical team used the targets from these generic models to optimize their respective subsystems. An update of the subsystem cost analyses will be presented by the Technical Teams at the November 1999 Collaborative Peer Review.

APPENDIX C

Presentations and Committee Activities

1. Site Visits to Ford Research Laboratories, Dearborn, Michigan; General Motors, Troy, Michigan; and DaimlerChrysler, Auburn Hills, Michigan, October 6–8, 1999.

2. Committee Meeting, November 17–18, 1999, USCAR Headquarters, Southfield, Michigan.

    Introductions
    Trevor Jones, Committee Chair
    Bernard Robertson, Sr. Vice President, Engineering Technology, DaimlerChrysler
    John Sargent (U.S. Department of Commerce), Chair, PNGV Government Technology Team

    Tier 2 Emissions Standards
    Charles Gray, EPA National Vehicle and Fuel Emissions Lab

    Discussion of Peer Review 5 Comments and Recommendations
    PNGV Executive Committee

    Specific Identification of Major Technical and Cost Barriers by Major Subsystems
    Al Murray, PNGV Technical Manager, Ford
    Chris Sloane, PNGV Technical Manager, General Motors
    Owen Viergutz, PNGV Technical Manager, DaimlerChrysler

Systems Analysis Review
Mutasim Salman, General Motors

Fuel Cell Technical Team Review
JoAnn Milliken, U.S. Department of Energy

Vehicle Engineering/Materials/Manufacturing and Vehicle Safety
Bill Stuef (Ford), Andy Sherman (Ford), Henry Thompson (General Motors)

Hard Trade-offs between Emissions and Fuel Economy
Ron York, PNGV Director, General Motors

Role of Fuels Industry in PNGV
Jim Spearot, Director Chemical and Environmental Sciences Laboratory, General Motors

Fuels/Propulsion Engines/Emissions Control Research
Michael Royce, DaimlerChrysler

4SDI Technical Team Review
Scott Low, Ford

Electrical/Electronics Technical Team Review
Bob Malcolm (DaimlerChrysler), Jim Merritt (DOE), David Hamilton (DOE) Balarama Murty (General Motors), Alex Gibson (Ford)

Battery Technical Team Review
Harold Haskins, Ford

Technology Transfer across the Vehicle Fleet
Steve Zimmer, PNGV Director, DaimlerChrysler

Strategies for Successful Migration to Production
Vince Fazio, PNGV Director, Ford

Summary of Highlights and Major Issues
Al Murray, PNGV Technical Manager, Ford
Chris Sloane, PNGV Technical Manager, General Motors
Owen Viergutz, PNGV Technical Manager, DaimlerChrysler

3. **Fuel Cell Subgroup and Battery Subgroup Meeting, USCAR Headquarters, Southfield, Michigan, January 12, 2000.**

4. Fuel Cell Subgroup Meetings, International Fuel Cells, Hartford, Connecticut, and General Motors Global Alternative Propulsion Center, Rochester, New York, February 2–3, 2000.

5. Committee Meeting, National Academy of Sciences, Washington, D.C., February 23–24, 2000.

Review of EPA R&D on Automotive Emissions Control Related to PNGV
Charles Gray, Jr., Director, Advanced Technology Division
EPA National Vehicle and Fuel Emissions Laboratory

PNGV and Tier 2 Emissions Regulations
Robert Perciasepe, EPA Assistant Administrator for Air and Radiation

Aluminum Body Approaches to Reducing Vehicle Weight
Richard Klimisch, VP Automotive, Aluminum Association, Inc.
Michael Wheeler, Automotive Consultant

Steel Body Approaches to Reducing Vehicle Weight
David C. Jeannes, Sr. Vice President, Market Development
Marcel van Schaik, Manager, Advanced Materials Technology American Iron and Steel Institute

Discussion of Concept Vehicles
Ron York, General Motors Corporation
Tom Moore, DaimlerChrysler Corporation
Arun Jaura and Andy Sherman, Ford Motor Company

Issues for PNGV beyond 2000
John Sargent, Chair, PNGV Government Technology Team,
U.S. Department of Commerce

Fuel Specifications for PNGV-Type Vehicles
Tom Bond, Manager, Fuels Technology (U.S.), BP Amoco

Overview of DOE Fuels Activities
Steven Chalk, Energy Conversion Team Leader,
U.S. Department of Energy

Adequacy and Balance of the PNGV Program
John Sargent (U.S. Department of Commerce), Chair
PNGV Government Technology Team

Appendix D

United States Council for Automotive Research Consortia

The U.S. automotive industry, through USCAR, has implemented collaborative projects that directly or indirectly support PNGV objectives. The USCAR consortia are listed below:

- Low Emissions Technologies R&D Partnership
- Automotive Materials Partnership
- Supercomputer Automotive Applications Partnership
- Natural Gas Vehicle Technology Partnership
- Advanced Battery Consortium
- Vehicle Recycling Partnership
- Auto/Oil Quality Improvement Research Program
- Environmental Research Consortium
- Low Emissions Paint Consortium
- Electrical Wiring Component Applications Partnership
- Automotive Composites Consortium
- Occupant Safety Research Partnership

Acronyms and Abbreviations

AC	alternating current
AEMD	automotive electric motor drive
AIPM	automotive integrated power module
ANL	Argonne National Laboratory
BIW	body-in-white
CARB	California Air Resources Board
CFRP	carbon fiber-reinforced polymer
CIDI	compression-ignition direct-injection
CO	carbon monoxide
CRADA	cooperative research and development agreement
DC	direct current
DECSE	Diesel Emission Control Sulfur Effects
DIATA	direct-injection, aluminum-block, through-bolt assembly
DOE	U.S. Department of Energy
EE Tech Team	Electrical and Electronic Power Systems Team
EGR	exhaust gas recirculation
EPA	Environmental Protection Agency
FEM	finite element method
GAPC	Global Alternative Propulsion Center

GM	General Motors Corporation
GFRP	glass fiber-reinforced polymer
HEV	hybrid-electric vehicle
HMO	hydrous metal oxide
IFC	International Fuel Cells
kWe	kilowatt electric
L	liter
LANL	Los Alamos National Laboratory
Li-ion	lithium ion
mpg	miles per gallon
NiMH	nickel metal hydride
NO_x	nitrogen oxides
ORNL	Oak Ridge National Laboratory
PEM	proton exchange membrane
PEVE	Panasonic Electric Vehicle Energy
PM	particulate matter
PNGV	Partnership for a New Generation of Vehicles
PNNL	Pacific Northwest National Laboratory
POX	partial oxidation
ppm	parts per million
R&D	research and development
SNL	Sandia National Laboratories
SO_2	sulfur dioxide
SPCO	Silicon Power Corporation
SUV	sport utility vehicle
ULEV	ultra low-emission vehicle
ULSAB	ultralight steel auto body
ULSAB-AVC	Ultralight Steel Auto Body-Advanced Vehicle Concepts
USCAR	United States Council for Automotive Research
VW	Volkswagen
4SDI	four-stroke direct-injection